W. W. (William Wellington) Greener

Modern Breech-Loaders

Sporting and Military

W. W. (William Wellington) Greener

Modern Breech-Loaders
Sporting and Military

ISBN/EAN: 9783741127069

Manufactured in Europe, USA, Canada, Australia, Japa

Cover: Foto ©Andreas Hilbeck / pixelio.de

Manufactured and distributed by brebook publishing software
(www.brebook.com)

W. W. (William Wellington) Greener

Modern Breech-Loaders

MODERN BREECH-LOADERS

Sporting and Military.

BY

W. W. GREENER,

SUCCESSOR TO WILLIAM GREENER, C.E.; AUTHOR OF "THE GUN," "GUNNERY IN 1858," ETC.

With Numerous Illustrations.

LONDON:

CASSELL, PETTER, AND GALPIN;

AND 596, BROADWAY, NEW YORK.

CONTENTS.

LIST OF ILLUSTRATIONS.

ILLUSTRATIONS OF SPORTING BREECH-LOADERS.

ILLUSTRATIONS OF REVOLVING AND DOUBLE-BARREL PISTOLS.

ILLUSTRATIONS OF MILITARY BREECH-LOADING RIFLES AND CARTRIDGES.

PREFACE.

In placing this volume in the hands of the reader, the Author feels it a duty he owes to the sporting and scientific world, to tender an apology for the disappointment occasioned by the non-appearance of "Gunnery in 1868," which was promised by his late father, W. Greener, C.E., whose death took place before the intended work was completed. The numerous inquiries made by subscribers to that work, and others wishing to possess it, cause the Author to feel it incumbent on him, as his father's successor, to publish this treatise on Modern Breech-loaders.

The important and numerous improvements which have recently been made in the construction of the breech-loader, sporting and military (especially the latter), have led to a great demand for information on the subject. This required information the Author has endeavoured to supply in the following work, in the production of which he has striven to condense within the smallest possible limits a full consideration of the various principles of breech-loading mechanism which he has found it necessary to introduce for description, and hopes, by the aid of numerous engravings and diagrams, to make himself clearly understood by the reader.

In all that appertains to the shooting and construction of the breech-loader, the Author has confined himself strictly to matters practical, and abstained from propounding any novel or startling theories. The remarks on "Black Gunpowder" introduced in this work, are reprinted from "Gunnery in 1858," as no important improvement in the manufacture of that substance has been made since that date, with the exception of the introduction of the so-called "Pebble Gunpowder," a full description of which compound has been given by

THE AUTHOR.

Modern Breech-Loaders.

TO THE READER.

The right-hand diagram of the first row at page 104 represents the Rigby form of rifling.

The diagram Fig. 86 (page 161) represents a section of a cartridge for revolver, 450 bore.

of accomplished workmen was brought to bear on the matter, breech-loading weapons were not favourably received for many years after that, which may be called a revival of the breech-loading principle. Mr. Lang, the well-known gun-maker, claims the merit of having introduced the new system to the notice of British sportsmen. The newly-invented breech-loading gun had, however, a powerful antagonist in the perfectly made and efficient English muzzle-loader of the period, which, by many of the most experienced sports-men, was pronounced to be all that need be desired. On

Modern Breech-Loaders.

The breech-loading system, although of very ancient origin, has been regarded with much doubt in this country, and, even in its modern and improved form, great reluctance has been shown to generally adopt it. This slowness of recognition and want of confidence may be said to be mainly owing to the adverse opinions entertained regarding it by both sportsmen and gun-makers. On the Continent matters were somewhat different, and the breech-loader achieved for itself a good reputation. It was, however, but little known in England until the Exhibition of 1851, when specimens of breech-loading arms were submitted by French gun-makers. Some considerable attention was directed at the time to these novel contrivances; but, notwithstanding that the skill of accomplished workmen was brought to bear on the matter, breech-loading weapons were not favourably received for many years after that, which may be called a revival of the breech-loading principle. Mr. Lang, the well-known gun-maker, claims the merit of having introduced the new system to the notice of British sportsmen. The newly-invented breech-loading gun had, however, a powerful antagonist in the perfectly made and efficient English muzzle-loader of the period, which, by many of the most experienced sportsmen, was pronounced to be all that need be desired. On the other hand, the shooting powers of the new breech-loader were imperfectly developed. The gun was to all intents and purposes French in design; it shot indifferently, and as yet the talent of English gun-makers had not overcome that serious objection, mainly because the opinion was general that the days of the breech-loader were numbered, and that it would not repay the trouble requisite to perfect it. At length, however, continental manufacturers, finding that their home demand continued to rapidly increase, turned their attention to the English market. Foreign breech-loaders were imported in large numbers, and then the gun-makers of England, awakened to the importance of immediate action, grappled with the difficulty in right good earnest, and how they have succeeded in overcoming it, we hope to show as we proceed.

To M. Lefaucheux is due the honour of inventing the modern breech-loading sporting gun; but although, so to speak, a practically useful weapon when first introduced by him, the action was weak and imperfectly developed. But his great achievement was the introduction of a shell or cartridge case which should fit the breech of the gun. This was a most important step towards the perfecting of the work he had so well begun. The shell or case, by expanding at the moment of discharge, effectually closes the breech joint and prevents the escape of gas. Conditions such as these had not been brought about before M. Lefaucheux's discovery, by the combined ingenuity of his predecessors.

The escape of gas was the great difficulty to be overcome, and however close the breech might be fitted the gas would, without the case, escape at the moment of firing, and find a way through the joints of the best-fitting breech; its doing so was owing to the expansion of the metal. The happy idea of making the cartridge contain its own ignition greatly contributed to the success of the invention. Strictly speaking, the cartridge case is the breech and nipple of the gun, the cap being inside the barrel; the brass striking pin becomes the nipple, thus the objection to conducting the flash from the outside to the inside of the barrel is overcome. The advantage gained by this is very considerable, as it does away with the escape of gas which takes place in a muzzle-loader through the nipple-hole.

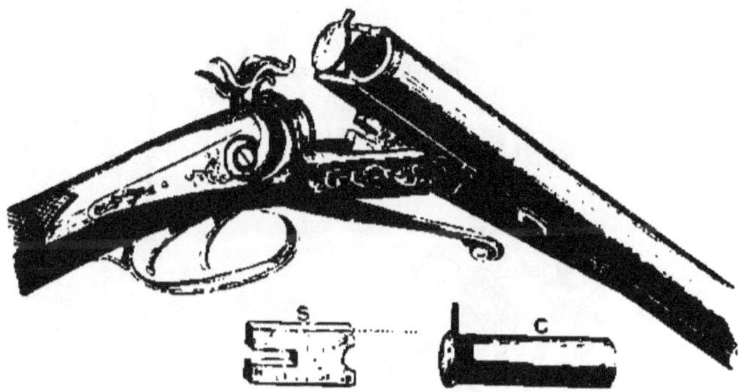

Fig. 1.—Lefaucheux Breech-loader. Single grip.

The pin cartridge case has been greatly improved upon by the rim-fire and central-fire principles, but these are only modifications of Lefaucheux's plan. (Invented in 1836.)

The great weakness in the first breech-loading arms that were introduced was the method of fastening the barrels to the stock. By referring to the engraving, Fig. 1, it will be seen that the breech-end is raised for loading. This is the most convenient and handy arrangement for double-barrel breech-loaders. Many other plans have been tried and abandoned. The "drop-down" system, as it is called, has many advantages: one is, that the barrels drop of themselves

Fig. 2.—The Sliding Action Breech-loader.

immediately they are disengaged by the lever; the hinge-pin effectually prevents the barrel from moving forward when secured in position for firing; the standing breech, which covers the breech-ends of the barrels, being solid, is a shield and protection to the shooter.

The face of this breech receives the force of the explosion; it is kept up in position by the strength or thickness of metal under the breech-ends of the barrels. The lever being required only to hold the barrels down, thus, the hinge-pin keeps the breech-ends of the barrels firmly up to the standing breech, when properly secured by the lever. It is very essential that the lever should be strong, so as to bind the barrels firmly to the action.

Lefaucheux's first gun had but a single grip, and this was about one inch from the hinge-pin, leaving that part unsecured that received the greatest force of the explosion, which is close to the breech. The result of this was that the breech sprang up a little every time the gun was discharged, and consequently made the barrels droop at the muzzle, which spoiled their shooting, besides causing a great escape of gas through the pin-hole.

This great defect was soon seen by English gun-makers, and improvements followed. Many methods were tried to remedy this evil, one being a plan called the sliding action. For loading this gun the barrels are pushed forward by a lever under the trigger guard, or in the front of the stock, as shown in the engraving, Fig. 2, about three inches from the breech, just sufficient to receive the cartridge case; the lever is then brought back into position, and fixes the barrels by a catch entering them underneath the stock, so as to prevent any forward motion at the moment of discharge. This plan was found to be insufficient to keep the barrels firmly against the false breech, and was discarded in consequence.

THE TURN-OVER ACTION.

This form of action was favourably received by many gun-makers, as it was considered to be free from several of the defects of the drop-down and sliding actions just described. It undoubtedly possessed more solidity than either of the be-fore-mentioned actions. The following is a description of it:—

Fig. 3.—The Turn-over Action Breech-loader.

The barrels are secured to the stock by a strong screw en-tering just below the extreme breech-ends; this screw also acts as a pivot on which the barrels turn for loading. This is accomplished by moving them to the right, describing part of a circle, and they are then in a position to receive the cartridges. When loaded, they are turned in the proper place for firing, and secured by a bolt entering the rib. This bolt works in the break-off between the hammers. This is the simplest arrangement of all the four principles described here, but did not meet with much patronage on account of several serious objections found to exist in it: viz., it was only suitable for the pin cart-ridge, and could not be made with bar or, as they are called, "*front-action locks;*" and besides, should by accident a tight-fitting case be inserted into the chamber, it would stop the working of the gun, as there is no leverage to open or close the breech.

THE SIDE-MOTION ACTION.

This plan of breech-loading is, in our opinion, the next best to the drop-down or Lefaucheux system. The plan is not new, although it has been recently patented by Mr. Jeffries, of Norwich. We have seen, many years ago, a single rifle made exactly on the same plan, evidently long before the Lefaucheux cartridge was introduced. The plan adopted for loading this rifle was to place the charge of powder and ball in a steel chamber about three inches long, having a nipple on the side. This tube was put into the barrels exactly in the same way as the Lefaucheux cartridge of the present day. The barrel moved on a pivot in the fore part of the stock, about nine inches from the false breech, to the right, for loading; and when in position for firing, it had

Fig. 4.—Jeffries' Side-motion Breech-loader.

the appearance of a muzzle-loader. It is only due to Mr. Jeffries to say that he has wonderfully improved on this plan. It will be seen by the sketch, Fig. 4, that the barrels turn sideways, and work on a broad steel plate, extending across the under part of their breech-ends, to which it is welded. This steel plate works in a dove-tailed slot cut across the action to correspond, both parts being tempered. In front of this there is what is termed a fork, working in the action; and the neck to which the guard-lever is fixed passes through the stock. This part opens the gun, and it is shut without the assistance of the lever, by merely closing the left hand, as quickly as any "*snap-gun.*" A stud is fixed under the forend of the barrels, which works in a slot cut in the forend of the action, enabling the barrels to move slightly forward, and allowing their end to clear the

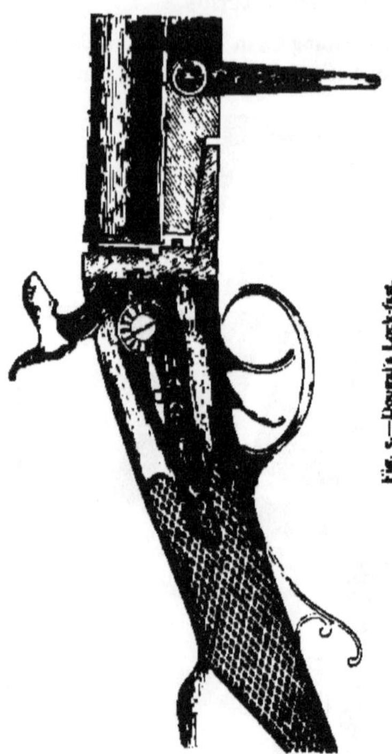

Fig. 5.—Dougal's Lock-fast.

face of the action. The extractor is an ingenious contrivance, and will answer for pin or central fire.

THE DOUGAL LOCK-FAST ACTION.

(*See* Fig. 5.)

This breech-loader partakes of two principles: it combines the Lefaucheux and sliding actions. It has the hinge-pin of the former and the disc of the latter. The hinge is an excentric rod connected with the lever. To disengage the barrels for loading, the lever is depressed. This moves the barrels forward about one-eighth of an inch, allowing the breech-ends to rise clear of the disc. When the cartridges are inserted, the barrels are held in position until the lever is brought back, which secures them for firing. The barrels are prevented from rising by the discs, and a steel lump placed underneath, which engages in the action.

To enable the reader to judge of the merits of the different systems of breech-loading double guns and rifles, we shall proceed to point out where the greatest strain is exerted, and which is the best kind of action to resist the force of the explosion. Now, in the first place, the standing breech which covers the ends of the barrels receives the greatest part of the explosive force. In all systems except Westley-Richards' this breech is kept up in position by the strength of metal under the barrels and at the angle. The first breech-loading guns introduced were made far too weak. They sprang open at the upper edge of the standing breech, there being no support except that given from the under side, and often broke through the angle at the proof-house. It was found necessary to strengthen these points materially. This was accomplished by leaving more metal at the angle, making it half-round instead of square, and rounding off the bottom edge of the barrels to correspond. The "Dougal" is an attempt to remedy this weakness by the disc and excentric, which force the barrels with extraordinary power against the standing breech, but fail to prevent the standing breech from springing back at the moment of discharge. In fact, it has no advantage over the modern Lefaucheux in this respect. The Dougal gun is generally made with back-action locks, which plan gives greater strength to the body or breech action, and this quality is most important for guns and rifles used with large charges of powder. We in all cases make our rifles with back-action locks, unless strictly ordered to the contrary, knowing as we do full well the necessity for having all the strength we can possibly get in the action. The bar locks necessitate the cutting away of the action to admit the mainsprings just under the barrels, thus weakening the body where strength is most required.

Fig. 6.—Wesley Richards' Patent Breech-loader.

Shot-guns made with bar locks answer very well, as

the strain is not nearly so great on the breech action when the shot is set into motion by degrees. But in the case of a bullet it is different. A projectile of this kind must be suddenly set in motion, which causes greater strain on the breech, and consequently more recoil. Our advice to all sportsmen who use large charges of powder, whether in rifles, ball-guns, or duck-guns, is to have their breech-loaders built with back-action locks. It is far more important to have the metal in the place where it is required to resist the strain, than to have any patent arrangement for locking only. It requires but a very moderate binding power to secure the barrels to the stock when the Lefaucheux hinge-pin is used; but it is very necessary to adopt a simple lever that will not work loose after a little use. Whatever plan is used, it should be made self-tightening, to compensate for wear and tear. The double grip is in our opinion the best suited for this purpose. It is an exceedingly simple and convenient arrangement. It has been thoroughly tried, having stood the test of many years' service, as applied to guns and rifles, and is generally approved of by sportsmen and all practical gun-makers. Dougal's patent action is deficient in this most important point, as when the excentric becomes a little worn the binding power is gone. But this can be guarded against by ordering extra excentrics to be sent a trifle larger, to replace the original when necessary.

Fig. 6 is constructed upon the most scientific plan, the barrels being secured to the breech action by an entirely different principle to all others. It binds the barrels at the top of the breech action. The hook which extends beyond the barrels from the rib is a kind of dovetail, which fits into the standing breech when in position for firing. The bolt or catch slides in the strap of the break-off, and engages in a slot at the top of the hook;

this bolt is pushed forward by a spring behind it, and is disengaged by pressing the lever to the right. The intention of this particular arrangement is to prevent the springing back of the standing breech at the moment of discharge. This is theoretically the best plan of securing the barrels to the false breech; but there are several objections to this system in practice. The barrels are secured by a spring bolt, which becomes loose after a time. The only remedy for this is to fit a new one. It is also open to the same objection as all other snap-action guns —viz., if a spring breaks it must be sent to the gunmaker for repair. We think this gun is better adapted for the pin cartridge than central fire. Westley-Richards' striker is made in two pieces working at different angles; the top striker receives the blow from the hammer like that of a pin cartridge, and communicates it to the bottom striker. There is no spring; the explosion of the cartridge forces it back; and the weapon has very much the appearance of a pin gun. Our wedge-fast breech-loader is similar to the Westley-Richards. Instead of a hook we substitute a steel lump with a round hole through it. (See Fig. 8.)

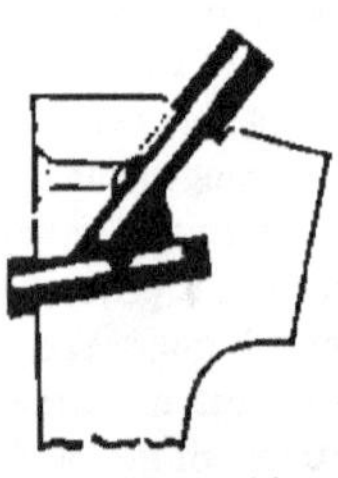

Fig. 7.—Westley-Richards' Patent Striker.

The barrels are disengaged by pressing the thumb against the lever towards the left-hand lock, moving the bolt in the opposite direction. The steel lump is a continuation of the rib firmly fixed between the barrels. The bolt passes through this, and effectually prevents the barrels from rising up or moving forward in the act of discharge; it also prevents the springing back of the standing breech. The sketch represents the bolt on the right side of the false breech; but it is more convenient to have it on the opposite side. The top-lever is a very handy mode of working the bolt in snap-action guns. The barrels can be disengaged for loading without taking the hand from the grip, only

Fig. 8.—Greener's Wedge-fast Breech-loader.

a slight pressure of the thumb of the right hand being required. Both these systems lack the necessary self-tightening power to secure durability. Wherever a spring

Fig. 9.—Top-lever Snap-action Breech-loader.

bolt is used the parts must be made very free, so that the barrels may be opened and shut easily. The result of this easy working is that the barrels soon become shaky, and rattle in the stock.

We have watched the working of guns constructed on the above system for several seasons, and find that they stand no better than the ordinary snap-guns. With the bolt that secures the barrels underneath, adjusted in the manner shown in the engraving (see Fig. 9), this gun has a much better and stronger spring than those above mentioned.

Fig. 9 represents a snap breech-loader, with a lever at the top and a bolt underneath the barrels. We have combined the two systems, and so dispense with the lump projecting beyond the barrels, which we consider inconvenient for loading, and, practically speaking, of no use. This arrangement we consider the neatest and most handy of all the snap actions.

PATENT SELF HALF-COCKING.

The barrels are disengaged by pressing the lever fitting over the trigger guard; the bolt works in the upper part of the break-off, which enters the barrel below the rib. By simply pressing the lever the barrels are disengaged, and the hammers raised to half-cock. The charges cannot explode without the barrels being securely locked; they can be loaded and fired with great rapidity; the arrangement is simple in construction, strong, and durable.

This form of breech-loader, which is represented at Fig. 10, has all the advantages of the snap-gun, but without a spring. The bolt which secures the barrel is kept in position by the hammers at the moment of explosion, so the self-cocking arrangement acts as a "safety."

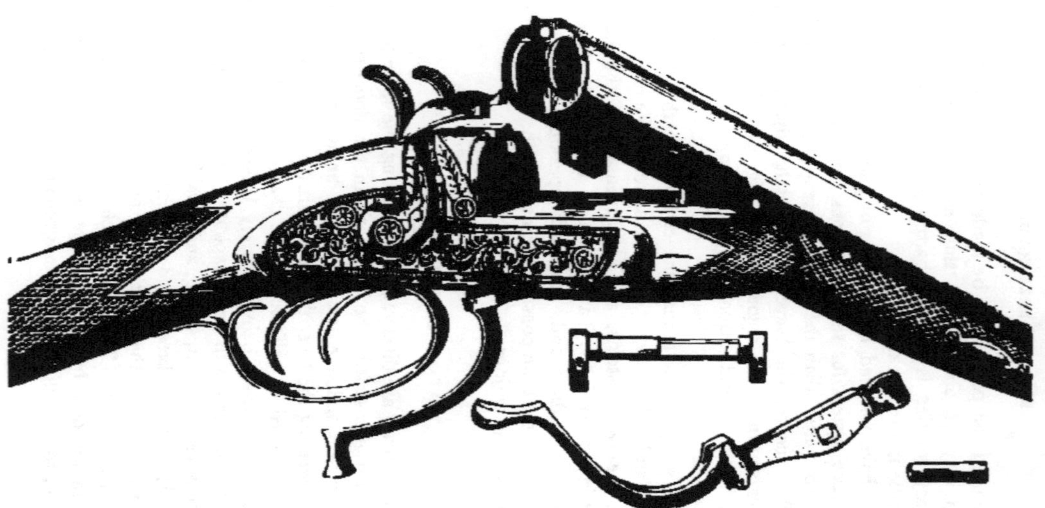

Fig. 10.—Greener's Self Half cocking Breech-loader.

We have made many large-bore double rifles on this system, and found them stand heavy charges of powder well, and give a very flat trajectory. All the wear is upon the bolt, which is easily repaired at a cost of a few shillings by fitting a new one composed of steel, that will make the gun as sound as when new. This gun is better adapted for pin than central fire.

POWELL'S PATENT SNAP ACTION.

This gun has the thumb-lever on the top of the break-off between the hammers. By raising the lever from the break-off the barrels are released and drop down, as in Lefaucheux's plan. The peculiarity of this action consists in the lever being so arranged as to form a catch which secures the barrels for firing; the lever is held in position by a spring, as shown at Fig. 11. The inventor of this gun considers the merit to lie in its extreme simplicity, but we are unable to detect any advantage in this arrangement over the ordinary snap-gun, except that the strikers are drawn back by raising the lever, no springs being required.

Such was the rage for snap-action breech-loaders in 1866-7, that all the principal gun-makers obtained patents for fancied improvements for securing the barrels, or arrangements of levers for withdrawing the bolt of the snap-action gun.

Some fixed the lever on the top of the break-off, some at the side of the lock, some underneath, and others at the fore part of the stock.

Many of these so-called improvements were simply contemptible, being only studs or thumb-bits, not worthy of the name of levers, requiring only a simple pressure of the thumb to disengage the barrels for loading. When these guns became foul it was difficult to get them open at all;

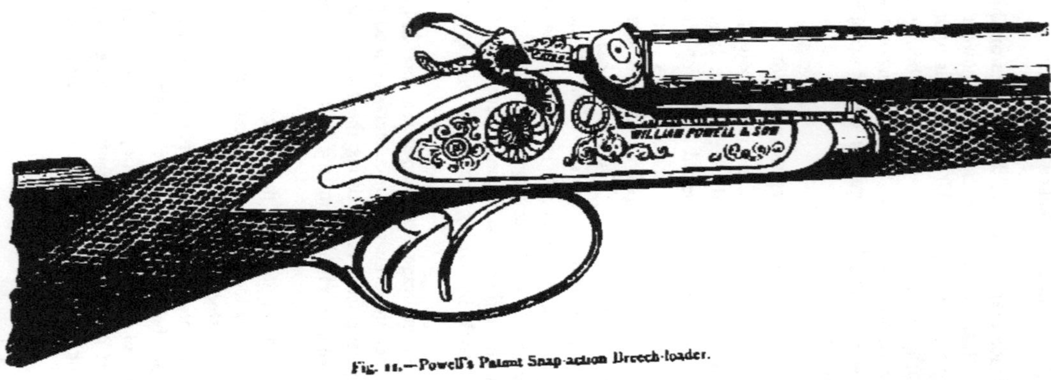

Fig. 11.—Powell's Patent Snap-action Breech-loader.

they had also a very bad habit of flying open at the critical moment of discharge. Several have been submitted to us for alteration to the double-grip plan. We have been told that they flew open repeatedly, but without injury to the shooter, the standing breech receiving the force of the discharge. The chance of injury to the operator by accidents of this kind is very small. We have seen barrels that have parted from the stock by the action breaking off short, close to the standing breech. Several cases have been reported in the sporting papers of accidents caused by the steel lump which secures the barrels to the stock giving way; in these cases the barrels have been hurled some distance without injury to the person firing. This is another proof of the safety of the Lefaucheux plan. Many breech-loaders have been made on the Continent with a movable breech arrangement, the barrels being fixed to the stock. This movable breech increases the risk to the sportsman. We consider them unsafe, for if anything gives way the head of the shooter is in danger. They have not the solidity of the standing breech. The double-grip is considered by all practical gun-makers to be the strongest and most durable arrangement for sporting guns and rifles. Nothing can be more simple or do the work better. There being a great amount of leverage, it possesses wonderful binding power, and when properly made and well fitted it will last many years without becoming loose, as it can be made self-tightening to allow for wear and tear. It is getting more into favour every season, amongst sportsmen at home and in India. Where guns and rifles constructed on the double-grip plan have failed, it is attributable to the imperfect mode of making the action. We have seen heavy double rifles of 8-bore fitted to a light breech action with bar locks, and the metal cut away just where it was most required, being left barely strong enough for a light-shot gun.

These imperfect guns are generally made by men who have no reputation at stake. They are sold by dealers, country gun-smiths, and even by some London makers so called. We have repeatedly pointed out these imperfections when guns of this class have come under our notice, and have always been met with the assertion that they have stood the proof, and that is all which is required. We are, however, of a different opinion: the commonest gun must stand the proof, a heavy fine being inflicted on all persons who evade it, provided detection follows; but this is no proof that the gun is well and perfectly made. It certainly may stand the strain at the proof, but then that is only one strain; it is the continual firing of heavy charges which tries the actions of large-bore rifles.

The metal used for making the bodies of breech-loading actions is of very excellent quality; it contains a considerable amount of steel: it is stiff and elastic. Soft iron would not answer. A light action made of this iron would resist the heavy proof charge, as the standing breech would spring back and then close up again. It is not the great charge of powder which causes such a formidable strain; it is the weight of the projectile and the spiral of the rifling; thus it follows that by adding to the weight of the bullet you increase the strain upon the breech-ends of the barrels and breech action; by adding to the charge of powder you greatly increase the strain upon the middles and muzzles of the barrels. Weak actions cause the barrels to droop at the muzzle, but we shall explain this more fully when describing the shooting of double rifles in a future chapter.

We shall now proceed to describe the different methods of attaching the steel lumps to the barrels, in order to secure them to the actions or stocks of breech-loaders. Several accidents have happened during the last season by the barrels parting company with the stocks while in the hands of sportsmen, and some very strong remarks have been from time to time made by correspondents to our leading sporting journals, relative to the great danger of purchasing guns with imperfectly brazed lumps. Now it is well known

No. 1.

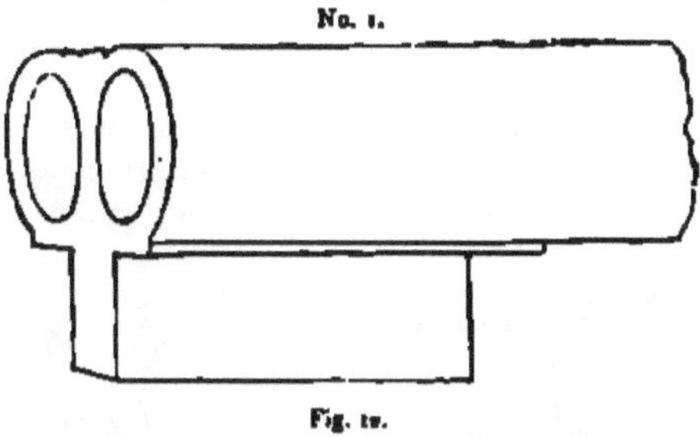

Fig. 12.

to all mechanics who are conversant with the process of brazing, that if two pieces of iron or steel are brazed together perfectly, they are virtually as strong as if they were

No. 2.

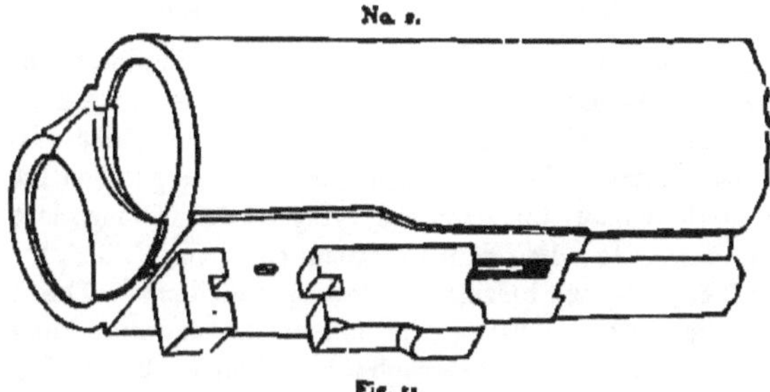

Fig. 13.

forged in one piece, especially if the pieces are small. The larger the surface required to be brazed, the greater the difficulty to get it sound; this is in consequence of the brass not running so effectually over a large surface as a small one. The engraving, Fig. 12, represents the breech-ends of the barrels with the lump brazed on in the first stage. It will be seen by referring to Fig. 13, No. 2, that the lump is considerably cut away to fit the stock; this is done by drilling, filing, and chipping with chisels, added to which the lump undergoes a great amount of hammering. This is the best test to prove if the parts are soundly brazed.

Fig. 14, No. 3, gives the end view of the steel lump brazed to the barrels; this simple V-shaped lump is the one most commonly used. There is no dovetail: it is

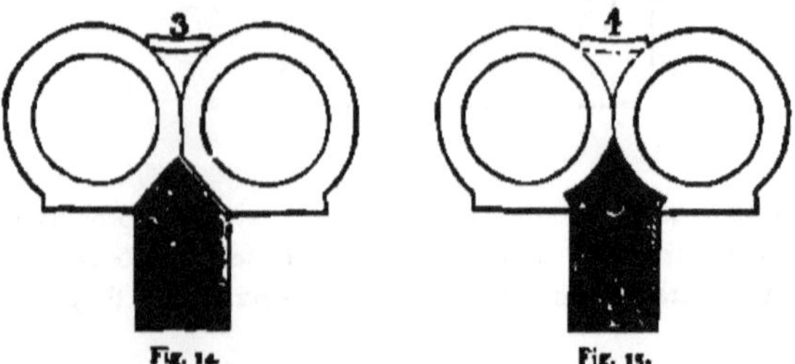

Fig. 14. Fig. 15.

entirely dependent upon the brazing for its adhesion to the barrels. We have made many hundreds of guns on this plan, and never found but one or two give way; and these mishaps occurred when we first began to make guns on the central-fire plan; the lump not being let deep enough into the barrels to admit of the extractor-hole being drilled, the drill cut away the brazing and loosened the lump. This is now remedied by having the hole for the extractor lower down, so that it may go through the solid lump. By referring to the engraving No. 2 the position of the extractor will be seen.

The engraving No. 4, Fig. 15, shows an improvement on the V-shaped lump, it being dovetailed as well as brazed. We consider this plan perfectly efficient for large-bore guns or rifles. If the brazing should happen to be imperfect, the dovetail would hold it so long as the barrels kept together. The chance of the barrels parting is very remote. The top rib being soft-soldered on is sufficient to keep them intact.

We have known this lump to stand without brazing at all—merely soft-soldered. There is certainly a slight chance of the dovetail being cut away a little, or weakened during the process of fitting the action; to obviate this, we have decided to use the lump as represented by engraving No. 5, Fig. 16.

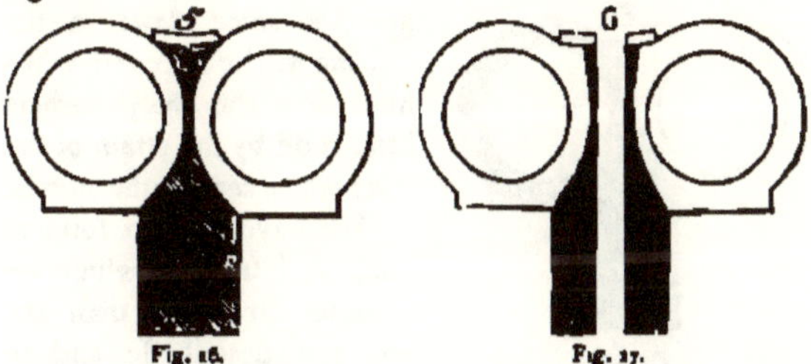

Fig. 16. Fig. 17.

This is a solid steel lump fitting between the barrels, extending upwards to the top rib; this lump can be brazed in the usual way, or soft-soldered. We consider if this lump is soft-soldered only, that it is perfectly safe. We have heard it argued that any dovetailed lump is liable to get loosened or come off by the barrels springing apart, supposing the brazing should be imperfect. We do not, however, concur in this view; but believe it next to impossible, simply because the barrels are so strong at the breech-ends, that they would not spring unless the barrels were loose at least twelve inches or more from the breech. This is an improbable condition of affairs, as the barrels are held together by the top and bottom ribs, besides being soldered between the two barrels. We prefer No. 5 lump to all others; it is, without exception, the best.

The engraving No. 6, Fig. 17, represents the lump patented by Mr. Parsons, of Birmingham. A piece of steel is welded to each barrel separately, as shown by the shaded lines; they are then made to fit accurately and brazed together, so forming a solid lump. The only weakness we perceive in this plan arises from its being brazed down the centre of the lump. As we have said before, there is a difficulty in getting a large surface perfectly brazed. In this form of lump, after being worked and shaped it is considerably weakened, and the chances are that some flaws in the brazing will be discovered. We have seen the sharp corners broken off by the strain of the lever which secures the barrels.

Fig. 18.

We have another form of lump to introduce, which we consider far better than the one just described; and to those who insist upon having a solid lump to the barrels of their breech-loaders, we recommend this form. (See No. 7, Fig. 18.)

We forge the lump on one barrel only—we prefer the right, as being more used than the left. We unite both barrels by a strong dovetail about three inches long. We secure them together by brazing or soft-soldering in the usual way. This leaves a solid steel lump to be worked upon, and none of the brazing is interfered with in the process of fitting the action, which is very liable to happen with Mr. Parsons' patent lump. There is no difficulty in welding a steel lump to a gun-barrel, as some suppose. We have used for many years barrels faced with steel, and have never known one to show a flaw.

CENTRAL-FIRE GUNS,

Now so universally adopted in this country, were introduced by Mr. Daw, the well-known London gun-maker. The cartridge has undergone very little alteration since its introduction by Mr. Daw. The principle is snap-action, with the lever over the trigger-guard; the barrels are disengaged by depressing the lever. This gun has the reputation of being strong and durable—it is usually made with back-action locks.

During the Exhibition of 1862 this gun attracted considerable notice in the sporting world. Its advantages over the pin-fire gun were maintained by many of the best shots of the day.

The greatest advantage gained by the central-fire principle is the non-escape of gas at the breech; the next is cleanliness; besides, there is no pin-hole in the barrels to let in the wet. This pin-hole is considered a great objection by some, as the pin must fit into the notch in the barrels before the barrels can be closed. In very rapid loading, and during excitement in battue shooting, or when after dangerous game in wild countries, this would, perhaps, cause delay in fitting the cartridge properly. "Delays are dangerous," especially when being charged by a bear or tiger. The central-fire plan greatly simplifies loading and unloading. It is often difficult to extract a tight-fitting cartridge from a pin gun; this is another cause of delay, especially when the gun is foul; besides, the cartridges are not so handy to carry, on account of the projecting pin, as the central-fire. On the other hand, pin-fire guns have advantages that central-fire have not; for instance, it can always be seen when they are loaded, because the pins of the cartridges stand up very prominently. This alone tells

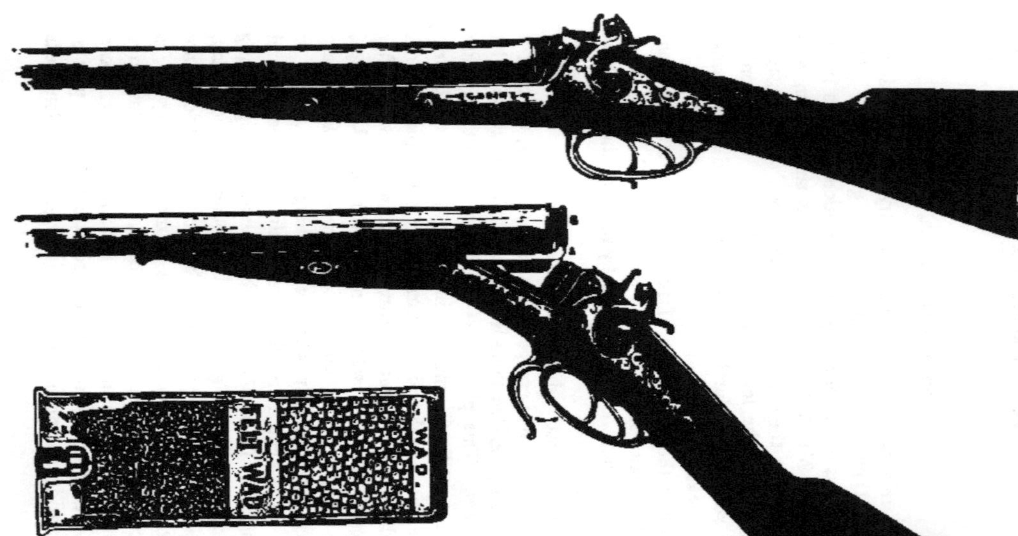

Fig. 20.—Daw's Central-fire Cartridge.

Fig. 19.—Daw's Central-fire Breech-loader.

wonderfully in favour of the principle, especially with very cautious sportsmen. It is more simple in construction than the central-fire, but the recent improvements in central-fire guns have been very great. In consequence of the demand for them, every means has been employed to perfect the system by gun-makers, and the demand is rapidly increasing. We already make a hundred central-fire guns to one "pin-fire."

The only objection some sportsmen have to the central-fire gun is that they cannot see at a glance if the gun is loaded. Certain gun-makers have patented a plan called an indicator, which consists of a pin fixed in the standing breech just over the striker, so that when a cartridge is in the barrel it forces the pin out. This arrangement was found very unsatisfactory—it is quite as easy to open the gun and ascertain if it is loaded, as to consult the indicator.

If every sportsman would observe the admirable rule of treating a gun as loaded until he has satisfied himself that it is not, the risk of accidental explosions would be reduced to a minimum. There cannot now be the slightest excuse for leaving a breech-loader with a loaded cartridge in it, on putting the gun aside when the day's sport is ended; and doing so should be considered an unpardonable offence in any case. One of the great advantages of the breech-loading system is that guns can be so readily loaded or unloaded, that if only a moderate amount of care were exercised, accidents with breech-loaders would become very rare.

Like all other inventions, the central-fire guns have undergone great improvements since their first introduction. The engraving Fig. 21 shows a modern central-fire breech-loader on the double-grip plan, having percussion fence and oblique spring strikers. The lever marked L is shown separate from the gun. This lever is secured to the gun by the screw and washer to a pivot passing through the lever,

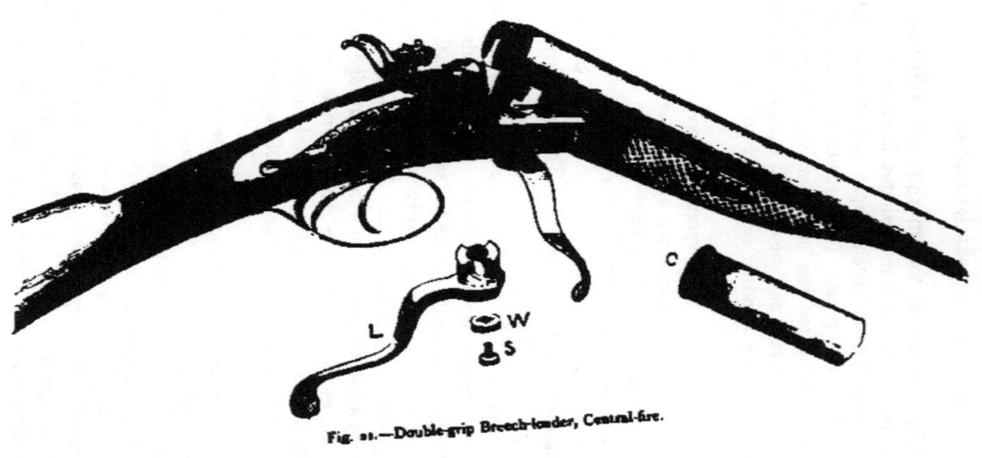

Fig. 21.—Double-grip Breech-loader, Central-fire.

the said pivot being solid with the action; there is a stop upon the washer which allows the lever to move one quarter of a circle. When the gun is closed the lever fits over the trigger guard. The cartridges are partially extracted by the act of opening the gun. To describe the form of extractor more minutely, we must refer to Fig. 13.

On examining the steel lump, it will be perceived that there is a small hole drilled through it length-ways, which extends from the breech-ends as far as the pivot on which the barrels turn for loading. The head of the extractor fits between the two barrels, and clips both cartridges by the rims, which are made larger for central-fire than for pin-fire guns. This prevents the extractor from slipping past the cartridge cases, and withdraws them about half an inch. The head of the extractor has a solid rod, attached, which passes through the hole in the lump as far as the joint; then it comes in contact with a projection on the joint, which forces it backwards when in the act of raising the barrels for loading. The weight of the barrels dropping down when open for loading is sufficient to extract the cartridge cases. If they should by chance stick fast in the chamber, they can be easily extracted by using a little pressure in forcing the gun open; this cannot by any means fail when the cartridge chambers are made the proper size. It is the most powerful extractor we are acquainted with; there is no military gun that possesses one half so good.

THE STRIKERS OR EXPLODING PINS FOR CENTRAL-FIRE BREECH-LOADERS.

The "Pottet" gun, as first introduced, had oblique strikers, similar to those represented in the last sketch.

I will endeavour to explain the working of these strikers very fully, and will introduce illustrations to make it better understood. The strikers are made of steel, with a collar in the centre, and kept in position by a nipple; the collar prevents it from coming out; it is allowed to play backwards and forwards freely. A small spiral spring is used to force the striker upwards, and keep it clear of the cartridge when closing the gun.

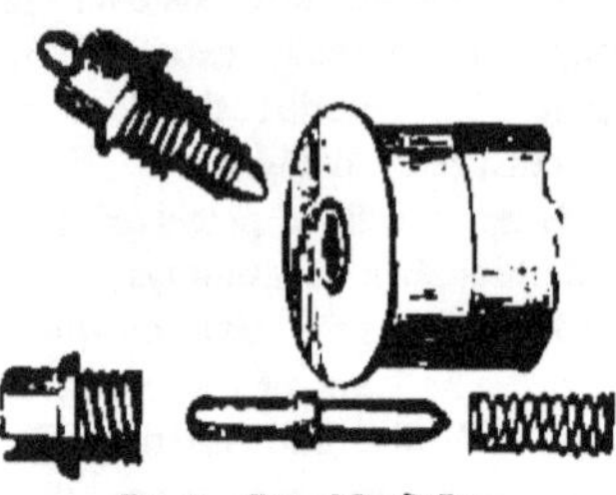

Fig. 22.—Central-fire Strikers.

We prefer the oblique to the direct striker, as here shown (Fig. 22); it admits of a longer hammer being used, a better blow is given, and the chance of a missfire is lessened. Many of the early central-fire guns were made with direct strikers, receiving the blow from the tumbler of the lock instead of the cock. This plan was found, upon trial, to be defective, on account of its constantly causing missfires; it lacked the swingeing blow that is given by the hammer to an oblique striker.

We wish now to point out a great danger that was experienced by many sportsmen who possessed guns with short strikers—and all breech-loaders were so made at first—and loud were the complaints made by sportsmen in the sporting papers of accidental discharges taking place when in the act of closing the gun. It was caused by the caps coming into sudden contact with the points of the strikers when the hammers were down. This was soon remedied by the very simple plan of making the strikers longer; this made the rim of the cartridge strike on the top of the point of the strikers, and so prevented the gun from closing at all until the hammers were drawn up to half-cock. It is best always to half-cock the gun after firing, and before attempting to reload.

REBOUNDING LOCKS.

This is a very simple and ingenious contrivance, invented to obviate the necessity of half-cocking. It is accomplished by lengthening the top part of the main-spring, and extending it towards the tumbler; the crank of the tumbler is lengthened beyond the swivel, and projects over the top part of the main-spring. At half-cock the crank of the tumbler rests upon the top part of the main-spring, and keeps the hammers from coming in contact with the strikers. This arrangement makes the lock partly self-acting; the hammers only require to be raised from half to full cock, instead of by the usual mode, which is from the nipple to half-cock, then to full cock. When in the latter position there is the same amount of force exerted by the main-spring upon the tumbler as in the ordinary locks. When the lock is released, by pulling the trigger, the hammer falls with enough force to drive the top part of the main-spring down sufficiently far to admit of the hammers striking the needles with power enough to explode the cartridge, when the hammers immediately rebound to half-cock. By the force of the top part of the main-spring acting upon the tumbler crank, and the striking pins being released, they rise clear of the cartridge. The gun can then be opened for reloading, without the necessity of half-cocking with the thumb. It is quite clear that this kind of lock cannot possess the same striking power as the ordinary lock, although it answers tolerably well when the caps of the cartridges are moderately sensitive. But we find that caps vary much in this respect—some require a good smart blow to explode them, in fact a much harder blow than can be given by the rebounding lock. To prevent missfires we find it necessary to use very strong main-springs, from 14 to 16 lbs. weight. Weaker main-springs will not explode some caps at all; we are convinced of this by experience, as we have been compelled to fit stronger main-springs to many of our early-made central-fire guns, and other makers have had to do the same. The necessity for doing so was caused by Eley Brothers making the caps much more sensitive at first than they are now doing. These sensitive caps in the cartridges were found to be rather dangerous, as in the act of closing the gun they have exploded, and in some instances with serious results. Eley Brothers very wisely remedied this defect by making the cap so that it would not explode without a powerful blow be given. Now, in the rebounding lock, if the lower part of the main-spring is made stronger to increase the striking power of the hammer, it follows that the top part of the main-spring must also be increased in strength, or the hammer would not rebound. We consider that sufficient striking power cannot be imparted to the rebounding lock to insure certainty of ignition with the cartridges as now made.

Another serious objection to the rebounding lock is that the strikers are not kept up to the cap during explosion; the consequence is that the caps are occasionally driven backwards into the needle-holes, causing an escape of gas, thus rusting and clogging the needles. This is very objectionable. We consider it very unsuitable for double rifles, where large charges of powder are used; the larger the charges of powder, the greater the escape of gas through the cap-hole. We have seen the cap forced up the striker-hole, thus stopping the working of the gun. It is very important that all double rifles intended for the pursuit of dangerous game should be constructed on such a plan as will prevent the chances of missfire as much as possible. They claim for the rebounding locks rapidity of loading, but this is a fallacy. To load quickly in a battue, or in cases of great excitement, the ordinary lock can be brought up to full cock at once; in fact both systems are on an equality as to rapidity. The advantage of the rebounding principle, in our opinion, has been greatly over-rated. It is on a false principle. Locks so constructed can easily be converted into the ordinary plan by simply fitting new main-springs.

SELF-ACTING NEEDLES OR STRIKERS FOR CENTRAL-FIRE GUNS.

A great number of central-fire guns are still made with spring strikers similar to those first used in the Pottet gun. Springs ought to be avoided as much as possible in the construction of fire-arms, especially the spiral spring, which is made of thin steel wire. It is a difficult matter to get really good spiral springs, that can be depended upon; they are sometimes too soft, and become useless in a very short time, or so hard that they break. They easily clog with oil and dirt, and then do not act at all; they are unsatisfactory at the best, even when kept well cleaned; they are rather dangerous if the cap should by any chance come into contact with the needles, when projecting from the false breech. We have seen these needles so fixed by being clogged with the escape of gas caused by the use of inferior cartridges, that it has been necessary to use an instrument to force them back. These matters may seem trifling, but they cause great annoyance to the sportsman. Gun-makers have long been sensible of this weakness, and many plans have been tried to work the striker with the hammer. Mr. Lancaster and other London makers, some years ago, tried a plan by which to raise the striker by the hammer, but did not succeed well. They made use of a direct striker engaging in the hammer near the tumbler. This plan withdraws the needles efficiently; but it fails in the most essential point, that is, the striking power. It is only a push that is given by the hammer to the striker; what is required is a "*blow*," and a good strong one it must be to prove effectual. Another weakness we must point out, and that is, that the cock does not hit fairly upon the head of the striker, but upon the arm or projection on the side; these arms are very liable to break off: such an accident would render the gun useless for the time. London makers have relinquished this plan some time, but many country makers are still having guns made on this principle. These weapons often come under our notice while being manufactured in Birmingham.

Mr. Horsley, a gun-maker of York, patented a plan by which to withdraw the needle or striker by the cock. He accomplishes it by a small lever or cam fixed in the false breech immediately in front of the breast of the cock. In raising the hammer to half-cock, the breast of the hammer comes in contact with the projecting end of the lever and pushes it forward; the corresponding half moves backwards, and brings out the strikers clear of the cartridges. This is a well-conceived plan, and gives a good fair blow to the cap of the cartridge. But it, however, has a weakness—the breech is too much cut away. Besides the hole for the striker to work in, a hole must be made for the lever, and another for the pin which fixes it. All this cutting away of metal in the breech is objectionable.

Mr. Pape, of Newcastle, has also a patent for a self-acting striker. It is not so simple or so good as Mr. Horsley's, and in our opinion there are too many parts in both these arrangements to answer well. Many gun-makers who have no patent of their own have appropriated the needle-rifle plan, with the needle fastened to the nose of the cock by a screw. This they can safely call a self-acting arrangement, and it fortunately belongs to any one who likes to make it.

We give a sketch of this German needle-gun striker, and point out the defects. The action partakes more of the character of a push than a blow. There is nothing to hold it in the false breech except the force of the main-spring; and should a bad cap be inserted, it would most probably blow away the striker. We have seen this result. It must also have a large hole at the back, or nipple, to allow the striker to work freely. This arrangement is bad; it allows the wet to get into the barrels; and the strikers frequently jar off close to the cock.

Fig. 23.—German Needle-gun Striker.

THE GREENER SELF-ACTING STRIKER.

It will be seen on reference to Fig. 24 that the striker is withdrawn by a small projection on the breast of the cock. The striker cannot turn round, as it is kept in position by a screw pin; this pin passes under the striker, which has a flat surface filed on the under-side—this answers for a groove. The top side of the screw pin keeps the striker in its place, allowing a piston-like motion. The hammer works the striker without being in any way attached to it. The hammer is raised half an inch when cocking the gun before the striker is moved. When the trigger is pulled, the cock falls quite half an inch before hitting the striker, thus allowing a good free blow to be given. That which we claim as a novelty in this invention is the withdrawal of the striker by the projecting arm, and the delivery by the striker of a fair blow on the head by the nose of the cock. All

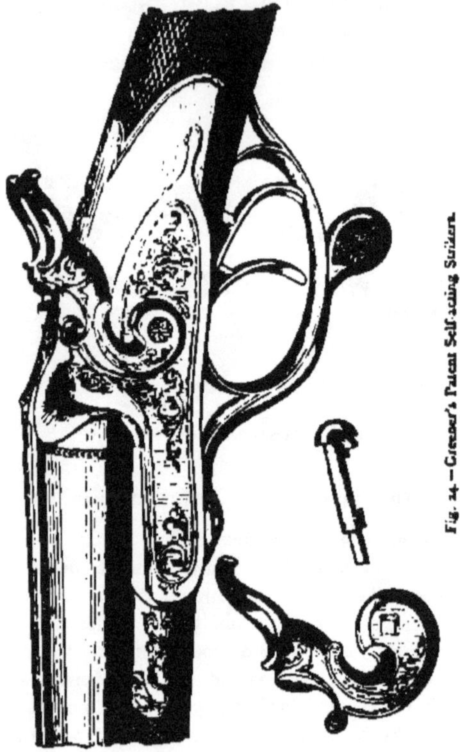

Fig. 24.—Greener's Patent Self-acting Strikers.

the mechanism can be seen at a glance. There is nothing concealed, no spring to get out of order, or anything that will impede the proper working of the gun. It is easy to clean, and works pleasantly, making the central-fire breech-loader all that could be desired even by the most fastidious sportsman.

There is another advantage we wish to point out. In our patent arrangement the striker being fitted so that it cannot turn round, it can be filed nearly flat at the point, so that the whole surface of the cap can receive a perfect blow,

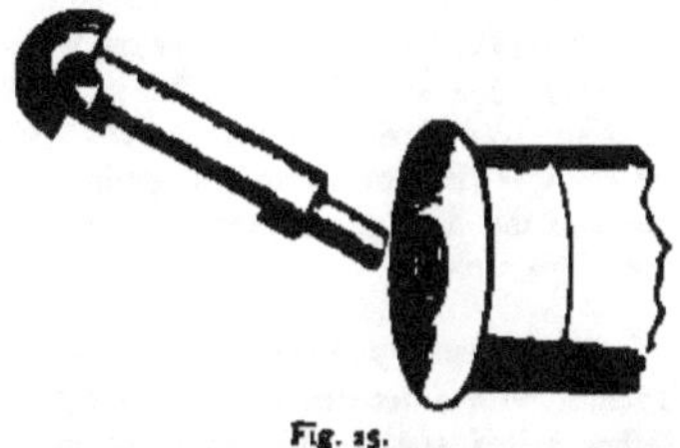

Fig. 25.

which insures ignition. This engraving represents the exact position of the striker and cartridge when in the gun. The spring strikers cannot be treated in this way, as they turn round, therefore must be made pointed. The anvil upon which the cap is exploded is pointed; the point of the anvil and the point of the striker cannot always be brought opposite; sometimes they strike the cap just over the point and sometimes under. Our striker is sure to find the point.

The following is the opinion given by three sporting papers, which will help to explain more clearly the advantages gained by this improvement:—

The Field.

We have repeatedly insisted on the danger attending any central-fire action in which the gun can be closed with the striker projecting from the false breech, especially if held there by the hammer. Several plans have been invented to meet this objection, and among the rest that illustrated below, which is patented by Mr. W. W. Greener, of Birmingham. It consists in connecting the striker and hammer by a pin and hook so arranged that at half-cock the striker is brought back sufficiently, yet is left so far from the hammer as to allow of a good blow being given.

Land and Water.

We notice an improvement in central-fire breech-loaders, patented by Mr. W. Greener, of St. Mary's Works, Birmingham, which we have examined, and believe to be a step in the right direction. We will endeavour to give a brief description of its merits. Instead of the strikers being worked by springs, as is usual in ordinary breech-loaders, he works them with the hammers, as will be seen by the sketch given. There is a small hook on the breast of the hammers, which bring up the strikers clear of the cartridge, at half-cock. The hammers are made to fall with a free and effective blow upon the strikers. We observe the needles are much stronger than usual, and altogether the gun appears neat and durable, and the arrangement very simple. We consider it worthy the notice of sportsmen.

The Museum of Firearms Journal.

From an inventor with antecedents and ancestry of the Greener stamp, whose father united considerable scientific with practical and literary ability, we naturally look for something above the ordinary run of patent productions. To many minds, a piece of complicated machinery with twice the number of parts and movements required for the accomplishment of a given object, is a splendid creation of inventive genius. Not so with the true inventor. He seeks to accomplish most by apparently small and simple means. The invention submitted to our notice follows in this track. The strikers (or, more properly, exploding pistons) of breech-loaders are usually held in position by small springs, which when the hammer is lifted raise the pistons. Sometimes the piston is left to work without springs, being lifted by the simple pressure of the cartridge. Both these features are objectionable, the spring being very liable to disarrangement and breakage; and without springs, the piston may clog, and stick in its socket. Mr. Greener's improvement makes it absolutely impossible for either one or the other contingency to arise. His pistons, by having a small arm at right angles, resting in a notch of the hammers, are lifted in the act of cocking, so that neither clogging nor spring-breaking can prevent the pistons from being taken out of the way. The engraving accompanying this notice will make the invention better understood, and the gun itself is on view at the Museum.

MISSFIRES WITH CENTRAL-FIRE CARTRIDGES.—I have read the correspondence on the subject of "missfires" in central-fire breech-loaders, and think the following may be acceptable :—Owing to conflicting statements, and perhaps, to some extent, to prejudice, I shot exclusively with muzzle-loaders up to a very late period. Having determined to give the breech-loaders a trial, I ordered a new gun of Mr. W. W. Greener, of Birmingham, on the principle advertised by him in *The Field*. I have used the gun at nearly all kinds of game, and in all kinds of weather. I am doubtful that I ever shot so well with a muzzle-loader as I have with the new gun; and I am confident I have never been so successful with "long shots." I have used Eley's cartridges, and I have not had a single missfire. The springs are, perhaps, stronger than usual. Both locks work equally well.—J. BURRKE (6, Alexandra Street, Westbourne Park, January 11). See *The Field*, January 16.

MISSFIRES IN CENTRAL-FIRE GUNS.—Sir,—In answer to J. F.'s letter in *The Field* of October 3, I beg to state that my subsequent experience of central-fire double guns quite confirms the statement that missfires are often owing to weakness of the main-springs of the locks. I have not had a single missfire since I have used a central-fire gun with W. W. Greener's new safety strikers, the locks of which had very strong and elastic main-springs.—W. (Deccan).

SHOOTING OF BREECH-LOADING SHOT-GUNS.

The first consideration in selecting a breech-loader is safety; the next is its shooting qualities. A large correspondence is carried on from time to time in the sporting papers on the subject of shooting, and varied are the accounts which continually appear. As to whether the breech-loader is equal in shooting power to the muzzle-loader, we have not the least hesitation in affirming that it is, when properly made by experienced and first-class makers who understand the secret of boring the barrels, &c. We shoot with all our breech-loading guns, and every first-class muzzle-loader that passes through our hands. We do this to satisfy ourselves, and keep a record of the shooting of all so tested for comparison with our own special breech-loader; and, we are proud to say, we can compete with any of them.

Many breech-loaders are made in Birmingham that never

did shoot well, and never can be made to do so. They are often ruined in the first stages of manufacture; they do not receive the care and attention that is necessary. The only object sought for is to get them up in large quantities for profit by the cheap makers; and if one or two should happen to shoot fairly, it is purely by accident. The usual size for breech-loaders is "*nominally*" No. 12 gauge: that means, to take the No. 12 cartridge case. But the size or bore of the barrel is left partly to the discretion of the maker, who bores it according to his own fancy. 13-bore is the "*actual*" size generally adopted; but some makers prefer 12 and even 11-bore. All these sizes cannot be adapted for correct shooting, as the inside of a cartridge case is exactly 11-bore. There is only one particular size that is suitable; and this has to be found out by repeated trials at a target. All first-class shooting gun-barrels will be found marked 13-bore. This mark is at the breech-end, stamped at the proof-house. In all cases where barrels are not bored up to the size before being proved, the proof-house people mark them the size under. For instance, supposing the barrels to be 13½-guage, they mark them No. 13; and again, if the No. 12 plug will not pass easily down the barrels the whole length, they still mark them No. 13. These marks are looked upon by some as denoting the exact bore of the gun, but this cannot always be depended upon.

All breech-loading barrels are fine-bored after they have received the proof-mark, in order to remove the indentations caused by stamping them at proof. It often happens that they are marked 13, and gauge full 12.

Some American sportsmen will insist upon having their guns marked 12 at the proof. This is a great mistake. It is impossible for us to make a really first-class shooting gun so marked. The same remarks apply to guns of 10-bore.

These should be marked 11-bore, which allows the barrels to be, when finished, just under 10-bore. We would strongly recommend all purchasers of breech-loaders to state the size of the cartridge case they wish to use, and leave the question of the bore of the barrel to the gun-maker.

The breech-loader is not bored in the same manner as the muzzle-loader. If the latter is bored true, it is of little consequence what the bore is; as it can be made to shoot well if the right charge of powder and shot is so selected as to suit the particular size of bore. The shot should lie compactly in the barrels in perfect layers. Whether it does so or not, is easily ascertained by putting a wad in the barrel about one-eighth of an inch from the muzzle, and putting on the top of the wad just as many shot of the right size as will fill up the bore in one perfect layer, so that there be no room left for half a pellet more.

In a breech-loader the above plan does not apply, as the charge is started in a No. 11-bore cartridge case—this is the exact size of a 12 cartridge case inside—and compressed into a 12-bore barrel. A certain amount of compression is necessary to obtain strong and close shooting. The penetration or pattern cannot be improved by increasing the amount of powder and shot beyond a certain charge. Superior shooting is dependent upon the comparative size of chamber and barrel, the method of boring, and the kind of metal the barrels are composed of.

Some gun-makers leave a nearly square shoulder from the chamber into the barrel. This is a bad arrangement, and causes great friction, unnecessary recoil, and also damages the shot. When properly made there is a gentle tapering from the chamber into the barrel. These are points which are carefully attended to by first-class makers; and without these conditions good shooting cannot be obtained. We find that three drams of moderately coarse grain powder, say

No. 3 Lawrence's, and 1⅛ oz. of No. 6 shot, is the very best charge for a No. 12 breech-loader. If the powder is increased no advantage is obtained; the shot is scattered, and no better penetration can be got by it. We have satisfied ourselves of this by repeated trials. If the charge of powder must be increased, let the shot be increased in proportion. An equal measure of powder and shot is best for any breech-loader of 16, 12, or 10-bore.

The result of using a larger charge of powder than is absolutely necessary is serious damage to the shot, through the extra force of powder driving the shot too suddenly against the taper of the chamber. The shot thus gets so malformed before leaving the barrels, that it spoils the shooting of a good gun. In this case, all the outer circle of shot is rendered useless over twenty yards. Eley's Wire Cartridges, that have such a good reputation for killing at long distances in muzzle-loaders, are almost useless in breech-loaders. The wire frame containing the shot is destroyed by passing from the cartridge case through the taper into the barrel. This wire frame is intended to keep the shot together for a certain distance and to increase the range, which it certainly does, so long as it keeps intact. It will often go for 100 yards without breaking at all; this is called "*balling*," but this happens only in a muzzle-loader. In a breech-loader the wire frame is driven in among the shot, and the fabric of the cartridge is entirely destroyed. Instead of being an advantage for long shots, it is a positive disadvantage. It scatters the shot more than when a loose charge is used. A few shots at a target will prove this statement to be correct.

It is quite clear that the form of chamber used in the ordinary breech-loader is detrimental to the use of wire cartridges, which the following account will illustrate:—

Some years ago, our father obtained a patent for what he called a "far-killing" breech-loader. His plan was shortening the cartridge case to about one-half the ordinary length, just sufficient to hold the charge of powder. The chamber of the gun was made to take this short cartridge only, which fitted up to a square shoulder. The charge of shot was made up in a separate cartridge, similar to Eley's Universal Cart-

Fig. 26.

ridge but without a wire cage, and made just to fit the bore of the gun. (See Fig. 26.)

One-fourth of the shot cartridge is placed in the cartridge case over the powder. The remaining three-fourths fit the bore of the gun, projecting beyond the chamber into the barrel. This plan ensures close and strong shooting equal to any muzzle-loader, with a shot cartridge; or a wire cartridge can be used in the same way. This plan answered too well, for many sportsmen sent back their guns to be altered to the regular form of arrangement. "They were indeed *far*-killing," for when used at close quarters they damaged the game to such an extent that our father discontinued making them, except for long distances. All duck-guns would be best made on the above plan. They would kill at a much longer range than ordinary guns, on account of the shot not becoming injured in passing from the chamber into the barrel, and will admit of a much larger charge of powder being used. This plan may be very easily tried in any ordinary breech-loader, by filling up the space over the powder with cotton wool, making up a separate cartridge for shot the exact size of the bore, and inserting it in the cartridge case and barrel as above described. By American sportsmen, who like to use large charges of powder for ducks, this will be found to increase the strength of shooting wonderfully. The objection raised to this plan is that when the right barrel is fired it displaces the shot in the left by concussion; but this need not happen if the precaution is taken to fasten it in by gum or some such adhesive material. It is the repeated firing of the right barrel which causes the displacement of the shot in the left barrel. Those who are in the habit of using one barrel more than the other should keep changing the cartridges, then no displacement of shot by concussion could take place. We have known cases where sportsmen have burst the barrels of their guns by firing four or five shots successively out of the right barrel while the left remained loaded. If the cartridges are not well turned down, the wad gets driven up the left barrel at every discharge of the right. We have known it move twelve or fifteen inches up the barrel. We point this out merely to put young sportsmen on their guard against accidents of this kind, although they may have learned their lesson in shooting and the proper way to handle the muzzle-loading gun. It must be remembered that it is the breech-loader we are dealing with now, which requires to be differently handled.

CONCENTRATORS.

Mr. Lancaster invented concentrators with a view to improve the shooting of breech-loaders. We have thoroughly tested them, and found that in some guns they improve the shooting quite 20 to 30 per cent. Mr. Lancaster says his concentrator will produce this effect in any breech-loader; but we differ in opinion, for the concentrator fired from some breech-loaders will injure the shooting rather than improve it. This is owing to the form of chamber; if the shoulder be too sudden it damages the form of concentrator. We have heard of a concentrator, so damaged, causing the bursting of the barrel of a breech-loader. It is supposed that the wad was forced past the concentrator, and left it in the barrel until the next shot was fired; such might have been the case, but we can scarcely believe that this was the real cause. We have only heard of one case of this kind. We think, however, a more suitable wad might be used—one that could not, by any means, pass through the concentrator; those sent out for this purpose being too thin. The wad should act as a perfect valve to prevent the gas from finding its way amongst the shot. The concentrator is merely a cylinder of paper, half an inch in length,

Fig. 27.—Concentrator.

and will just fit in a No. 12 cartridge case (see Fig. 27). To load the cartridge case, either pin or central-fire, load with usual charge of powder, putting over it first a thin card or mill-board, and then a best white greased cloth wadding.

If loading with Erskine's machine, after ramming the wadding down on the powder, first insert the concentrator in the mouth of the case; put in the shot, then the usual cloth wadding, and ram well down, which will drive the concentrator home into its proper position, and finish by turning over the cartridge as usual. If employing any of the ordinary machines for loading one cartridge at a time, the concentrator may be put in before or after the shot has been placed in the case, at the option of the loader; care must however be taken that the concentrator, when home, is in the position shown in the illustration, that is, level with the top of the shot.

Any one wishing to try these concentrators, and not

having the proper ones handy, can, by cutting about half an inch off the end of a 16-bore cartridge case, make a very good substitute; this will just fit a 12-bore cartridge case. We think that these concentrators might be tried, and perhaps with advantage, in guns that do not shoot the regular charge well; and gentlemen having ball-guns in India might improve them for shot by using the said concentrators; at any rate it might be worth while to give them a trial; but we do not see any advantage in using them in first-class shooting guns—that is, in guns that will put 130 pellets in a 30-inch round target at forty yards, and penetrate thirty sheets of thick brown paper with a charge of three drams of powder and 1⅛ oz. No. 6 shot. Such guns as these need no improving, they are all that is necessary. Our best guns are made to do this. We repeatedly hear from our customers that with them they can kill grouse at sixty yards, and old cock pheasants at fifty yards.

CONVERTING MUZZLE-LOADERS INTO BREECH-LOADERS.

It is often a difficult matter for the gun-maker to answer off-hand the inquiries repeatedly made relative to alteration by sportsmen who have favourite muzzle-loaders, and wish to have them converted into breech-loaders. Some guns will admit of conversion, but others will not, and in the latter case it is only a waste of money to attempt the operation.

To save time, and the vexation arising from having a good old muzzle-loader spoiled by being converted, we will point out what kind of barrels are best adapted for converting. In the first place, they should be very strong at the breech, to allow for boring the chambers, so as to admit the cartridge case, and yet be strong enough to resist the large proof charge that they are subjected to. There are but few muzzle-loaders strong enough for converting. If the guns are wanted to retain their good shooting qualities, they must be fitted with cartridges one size larger than the bore or gauge at the muzzle, if the bore is over 13; but not for a 16-bore, as they take a 15 wad. It should be a No. 14 cartridge for a 15-bore gun, and a No. 12 for a 13-bore.

If the above conditions are complied with, the guns, if properly converted, will shoot as well, and we have known them shoot better than before alteration.

There being no 11-bore cartridge cases made, 12-bore guns cannot be converted, to retain their good shooting qualities. An 11-bore gun can be made to take a 10-bore cartridge case. All converted guns have to undergo the process of double proving, and require to be fine bored, which increases the size.

There has just been invented by an American a metallic cartridge case, that will answer well for 12-bore converted guns, and will, no doubt, become general and have a large sale. These cases are thinner than the ordinary paper case, and fit well into the 12-chamber. Being less substance than paper, they take a wad of 11-bore, and can be used repeatedly.

CONVERTING MUZZLE-LOADING RIFLES INTO BREECH-LOADERS.

This is a delicate matter, and should be well thought over, and the rifle examined, before subjecting it to the ordeal of conversion. There are more difficulties connected with the alteration of a rifle than are encountered in converting a smooth-bore. For instance, a spherical-ball, muzzle-loading rifle takes a bullet with a patch, fitting just easy enough to be forced down with the ramrod. The very reverse is the case with a breech-loading rifle—the bullet must be at least one size larger than the bore, so that the projectile may fill up all the grooves of the rifling without a patch. This is absolutely necessary to insure good shooting. The bottom of the grooves is, correctly, the bore or gauge of the barrel in a breech-loading rifle. For instance, a 12-bore cartridge case will take a No. 11-bore bullet, fitting it tightly. It is usual to make all 12-bore rifles to take the 11-bore bullet, so that a 12-bore breech-loading rifle takes a bullet the same size as an 11-bore muzzle-loader. An ordinary 12-bore muzzle-loader will convert into a breech-loader, provided that the grooves are not cut too deep; but it would require to be regulated again for shooting.

Rifles of 14 or 16-gauge will convert in the same manner—that is, taking cartridge cases of 14 or 16, and a bullet 13 or 15. All Eley's cartridge cases take a bullet one size larger than they are marked. The rifling in a muzzle-loader is not exactly the pattern that we put in breech-loaders—the latter answers better with shallower and wider grooves. It is impossible for a bullet to strip in a breech-loader, if properly fitted, even if used with a very large charge of powder. This is one of the great advantages that the breech-loading rifle has over the muzzle-loader.

The Enfield rifle 577-bore will convert, and make a good accurate breech-loader. The process is made simple by using the Boxer case—thanks to the Government, for they have spared no trouble or expense to adapt the Boxer cartridge to the regulation rifle; they have been able to overcome difficulties that no private maker could have surmounted.

These rifles will also shoot well with a spherical bullet (twenty-three to the pound), and admit of a large charge of powder being used.

THE BACON BREECH-LOADER.

We introduce this action as being the only one we are acquainted with on the bolt principle adapted for double guns constructed after the style of the Prussian needle-gun. We give the Editor of *The Field's* opinion of it, and also a representation of the gun from the block made use of to illustrate the two letters here given, which appeared shortly after the editorial notice in that valuable paper.

Editor's Notice.

MR. BACON'S NEW BREECH-LOADER.—At length we have presented to our notice a breech-loader loading each barrel separately, yet equally quick in its action with the snap-action hinged gun, and presenting neither snap nor hinge. This combination of rapidity with the absence of snap is effected by a provision which completely extracts the empty case, and drops it away from the gun. In order, however, to explain its mechanism, we must have recourse to the engravings.

Fig. 1 shows the gun closed, the left barrel being unloaded with its striker down, and the right loaded and unbolted, the bolt answering the purpose of half-cocking. The barrels are each utilised for the reception of the false breech, in which is the lock, the striker being driven by a spiral spring. When reloading, the lever is laid hold of and turned up sharply, after which it is drawn back, as shown in Fig. 2, when it drops the empty case through a slot in the under surface of the space between the barrels, and then through the stock. A fresh case is then inserted, pushed forward, and the lever depressed to its original position, where it is firmly held by two cams of the ordinary description. A similar process loads the other barrel; but so quickly are these operations effected, that we have seen eighteen shots fired in the minute. The ordinary central-fire cartridge case is used, and, as far as we have seen, both the extraction and the ignition have been perfectly performed.

The only objections that occur to us are the novelty, and possibly it may be said the unsightliness, of the gun, together with the addition of about 2½ in. to the length of the barrels. We can offer no opinion as to the durability of the action, because we have not had time to test it; but the wearing surfaces are large, and even if worn a good deal, there would be no loss of strength. The plan is, no doubt, sound and cheap; while its handiness is quite first-class, and the facility of re-loading one barrel, and of firing the other while in the act of loading, are great recommendations. Mr. Bacon, who is a retired officer in the Marine Artillery, deserves great credit for his invention, and we can recommend our readers to give it a trial with full confidence in its merits, which closely approach, if they do not surpass, the best actions of the day. The gun is highly spoken of by some of our subscribers, and may be seen at *The Field* office.

THE BACON BREECH-LOADER.—Having had recently an opportunity of inspecting one of these guns, I think it right to draw attention to one point overlooked in your notice, and, indeed, in every notice of the gun which I have seen. Before deciding whether the manipulation of this gun is without danger to the shooter, it will be requisite to ascertain what is the probability of a cartridge being exploded in the act of shooting the gun. The special committee on small arms came to the conclusion that this was a source of danger in all guns closed by thrusting forward a bolt in rear of the cartridge, and rejected all the "bolt" systems partly on this account. In closing the Bacon gun, it seemed to me that the bolt delivered a rather smart blow on the base of the cartridge. This, with a very sensitive cartridge, would probably cause an explosion, the result of which would be to drive the bolt into the hand of the shooter. It is desirable that this point should be clearly settled before the Bacon or any other "bolt" arrangement is adopted as a substitute for the present system of double guns, in which the explosion of a cartridge while closing the action is unattended with danger to the shooter. SAFETY.

[There is no doubt that, as the striker of the Bacon gun is now made, a sensitive cap might be exploded; but it is extremely easy to make the hole through which the head of the striker passes as large as the cap, and in that case there would be even less danger of igniting a sensitive cap than in any breech block which rises or falls, as in the latter case a "proud" cap, if sensitive, may be exploded.—ED.]

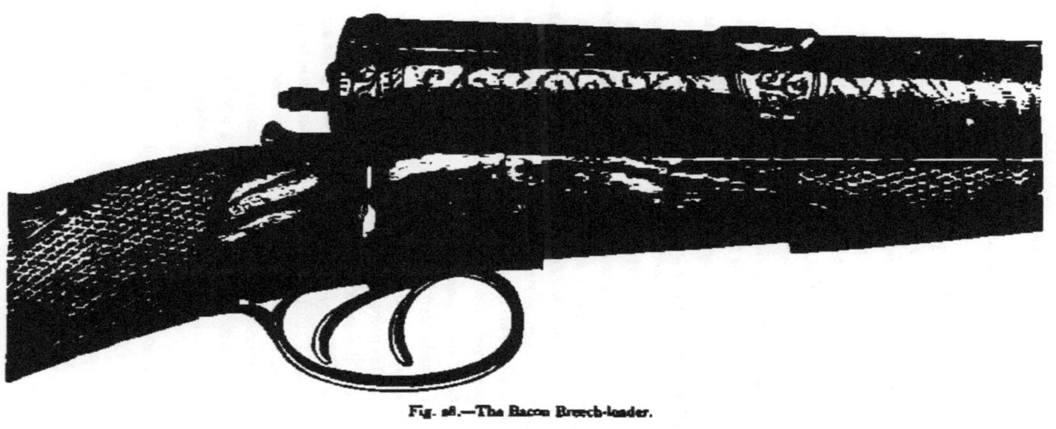

Fig. 28.—The Bacon Breech-loader.

THE BACON BREECH-LOADER.—Sir,—You have selected with such discrimination the point which conclusively answers in fewest words the objection raised to my gun by "Safety" in last week's *Field*, that it is only necessary for me to say, that being prepared for this prejudice, I ordered the manufacturers at the close of last season to make a depression in the face of the so-called "bolt" in rear of the cartridge cap, thus adopting by anticipation your identical suggestion. I wish, however, to state that I did this simply to remove all possible grounds for prejudice, believing the gun to be perfectly safe without it. The breech face does not deliver a blow on the cartridge base as "Safety" supposes, the blow being taken by the face of the flange encircling the rear end of the action, except in the single instance where the cartridge may have a flange too thick for its recess in the barrel, and even then the extractor acts as a "buffer spring" against the blow; whereas in a Lefaucheux gun the base of such a cartridge receives a clean slap against the false breech, delivered with the aid of a powerful lever in closing the action.

In proof of my own confidence, I appeal to the fact that I have fired 9,000 rounds from actions similarly constructed with respect to the point in question, and have never caused a cap to explode by closing the action; nor has that occurred even once in the innumerable instances in which I have exhibited the action with capped cases, nor amongst the many thousands of rounds fired last season from these guns by other people.

The report of the Committee alluded to by "Safety" was most satisfactory on this point. In all cases where they exploded caps with "bolt" actions, the mark of the striker was found upon the cap; whereas they failed to cause an explosion in closing the action when the striker was so constructed that it could not protrude in doing so, although they tested "bolt" actions with caps made specially sensitive. In the report this is stated with all fairness, and the ground of objection on this head mainly rests on the fact of a cartridge, out of the vast numbers manufactured, having been known to explode from falling on the floor of a factory, which is clearly not a parallel case, because it is impossible to be sure that grit or other excrescence was not at the point of contact. FRANCIS BACON.

Wymondham Rectory, Oakham, April 18.

Fig. 29.—The Bacon Breech-loader.

THE HORSLEY BREECH-LOADER.

The accompanying diagram represents a patent breech-loader invented by Mr. Horsley, of York. It is of the form of construction known as snap-action. The lever for opening and shutting the gun is situated upon the strap of the break-off, and is drawn backwards by the thumb of the right hand to disengage the barrels—a different motion to any other patent described in this work. The bolt that secures the barrels for firing enters the steel lump below the barrels, the same as in ordinary snap-action guns. It works pleasantly, and has an advantage over many other snap-guns; for should the spring break, it would not render the gun unserviceable, as the lever could be kept in its place by the thumb at the moment of discharge. There is another novelty in this patent—the strikers are withdrawn by the hammers in a very ingenious way, there being a small lever or cam fixed in the false breech in front of the breast of the cock; in raising the hammers to half-cock, the breast of the cock comes in contact with the projecting end of the lever, and pushes it forward; the corresponding half moves backward and brings out the strikers clear of the cartridges. The only weakness we can detect in this arrangement is that the breech is cut away far too much.

There is an indicator to show when the gun is loaded. This is the best arrangement of the sort we have seen, as the letters are in gold, and show very plainly the word *loaded*. There is a sliding shield on the top of the false breech, which has a peg projecting from the face of the standing breech. When the cartridge is inserted in the chamber, the peg is forced back, removing the shield, and showing the word "loaded." We, however, consider all this arrangement superfluous when in the hands of a careful

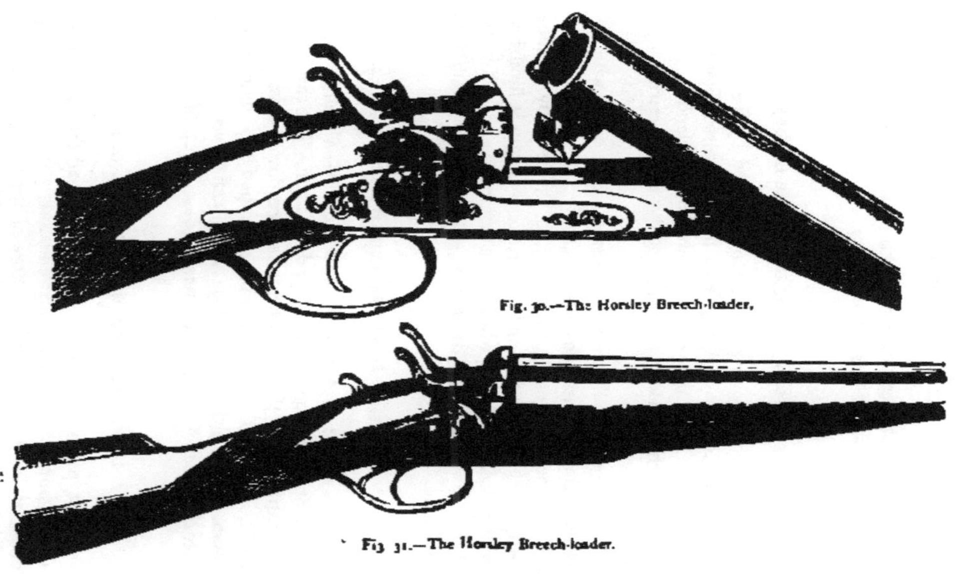

Fig. 30.—The Horsley Breech-loader.

Fig. 31.—The Horsley Breech-loader.

sportsman, but there are many who would like the principle and mechanism of this patent gun.

THE WESSON AMERICAN BREECH-LOADER.

This breech-loading gun is manufactured by the Wesson Firearms Company, Springfield, Mass., United States of America. It is not much known in England. It is partly machine-made, and thoroughly an American style of gun.

The annexed illustration shows the gun open, with the stock, hammer, and side-plate removed.

To open the gun for loading, raise the lever by pressing with the thumb upward and forward. The gun is furnished with rebounding locks, whereby the hammers, after delivering their blows upon the firing pins, return to the position of half-cock.

THE ALLEN BREECH-LOADER.

This is also an American breech-loader—a modification of the Snider action, adapted to a double gun. The cartridges are extracted by the guard being pushed downwards. Steel shells are used, which can be re-capped any number of times. The ordinary cartridge cases can be used in this breech-loader. We give the maker's description of it.

DIRECTIONS FOR WORKING THE GUN.

First, put on the cap or primer, then load the steel shells as you would an ordinary muzzle-loading gun—always using one size larger wad than the bore of the shell; then open the lid of the gun and insert the cartridge; then close the lid, and it is ready to cock and fire. After discharging,

Fig. 32.—The Wesson American Breech-loader.

open the lid and drop the guard, to withdraw the shells, and the operation is complete. The wad-driver or loader is also used to force out the primer from the shells, by first turning the ferrule on the loader to the left, then slide forward, then to the left again; and finally force out the primer. By turning off the knob—the loader—you have a worm for drawing charges. To detach the stock from the barrels, open the lid and remove the guard and the large screw in front of the guard, and slide the barrels off laterally.

Care should be taken to keep the cartridges well oiled, and the gun cleaned after the day's shooting.

☞ *A soft felt wad put upon the charge well saturated with oil, and occasionally shot from the gun, will clean the barrels wonderfully.*

THE ABBEY BREECH-LOADER.

This is another American breech-loader, invented by Mr. Abbey, of Chicago. It is an adaptation or modification of certain well-known patents—one of these is Purday's Double Bolt, with the lever over the trigger-guard; with a vertical bolt that slides up the standing breech, and engages in a steel lump which is a continuation of the top rib—similar to the Westley-Richards plan. This vertical bolt is not new, as a gun on this plan was exhibited at the "Field" Gun Trial of 1866, without the addition of the bottom bolt. All that can be claimed by Mr. Abbey is the combination of the above-mentioned principles. It is maintained by the inventor that this gun cannot get wide at the joint. It is much better for the maker to add more metal where it is necessary, which would answer quite as well, and have the advantage of being more simple in construction. We have tried many experiments to satisfy ourselves on this point, and we feel sure there is no springing back of the breech

Fig. 33.—The Allen American Breech-loader.

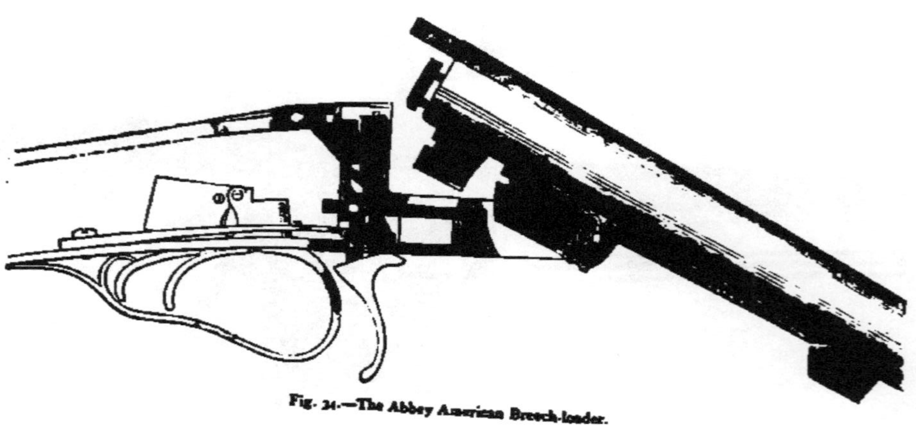

Fig. 24.—The Abbey American Breech-loader.

1866.

RESULT of t[…]rms," Old Brompton, on the 22nd and 23rd of May, 1866. All the shots […]r plate of iron, 30 inches in diameter, having a square of paper su[…]sses of double-imperial brown paper, 140lb. per ream), procured fr[…], the size being 10½in. by 9½in.—in round numbers, 10 inches s[…] counting the penetration the number of sheets broken by any one […]llets per ounce. Powder—Curtis and Harvey's No. 3, 5, or 6, at t[…]at the eight guns highest on the list used No. 5 or 6, and the se[…]

25. Mr. Hasdell......	68	92	87.4	Right.. 23	19	22	25	27	17	22.1	22.2	220
	56	83.2		Left ... 26	17	22	[illegible]	26	[illegible]	22.3		
26. Cogswell & Har[…]	96	75.2	83.2	Right.. 30	25	25	24	21	22	24.3	26.[illegible]	219
	111	91.2		Left ... 24	26	33	26	25	33	27.5		
27. Mr. Joynson ...	41	79.2	84.4	Right.. 28	27	27	25	22	19	24.4	24.1	217.4
	96	90		Left ... 23	24	28	20	23	24	23.4		
28. Mr. Joynson ...	94	87	83.2	Right.. 36	23	26	28	22	32	27.5	25.1	217
	76	79.4		Left ... 19	25	22	24	21	24	22.3		
29. Mr. Fletcher ...	86	85.3	80.2	Right.. 19	20	28	27	35	20	24.5	25	210.4
	57	75.2		Left ... 20	22	27	22	26	27	25		
30. Mr. Hasdell......	106	79.1	75.5	Right.. 27	27	27	31	26	24	27	27.1	205.5
	84	72.3		Left ... 31	30	30	26	23	23	27.1		
31. Mr. Tolley	81	71.3	69.4	Right.. 32	13	18	20	27	21	21.5	20.4	180.4
	76	68		Left ... 26	18	21	13	19	29	19.2		
32. Mr. Hast	57	67.1	62.1	Right.. 32	23	15	23	17	15	20.5	20	164.2
	54	57.1		Left ... 24	20	21	18	15	17	19.1		
	....		98	Mean Average of 32 guns							24.1	244.2

C[…]

1. Mr. Pape	69	100.3	109.4	Right.. 15	29	19	24	20	20	21.1	22.5	264.5
	88	118.4		Left ... 22	23	26	26	28	22	24.3		
2. Mr. Elliott	97	78.1	73	Right.. 21	22	22	20	23	23	21.5	21.4	182.4
	78	67.5		Left ... 18	21	26	17	20	23	20.5		

C[…]

1. Mr. Elliott	107	111	108.2	Right.. 27	23	27	24	25	38	27.2	28	277
	148	105.5		Left ... 24	29	23	30	31	34	28.3		

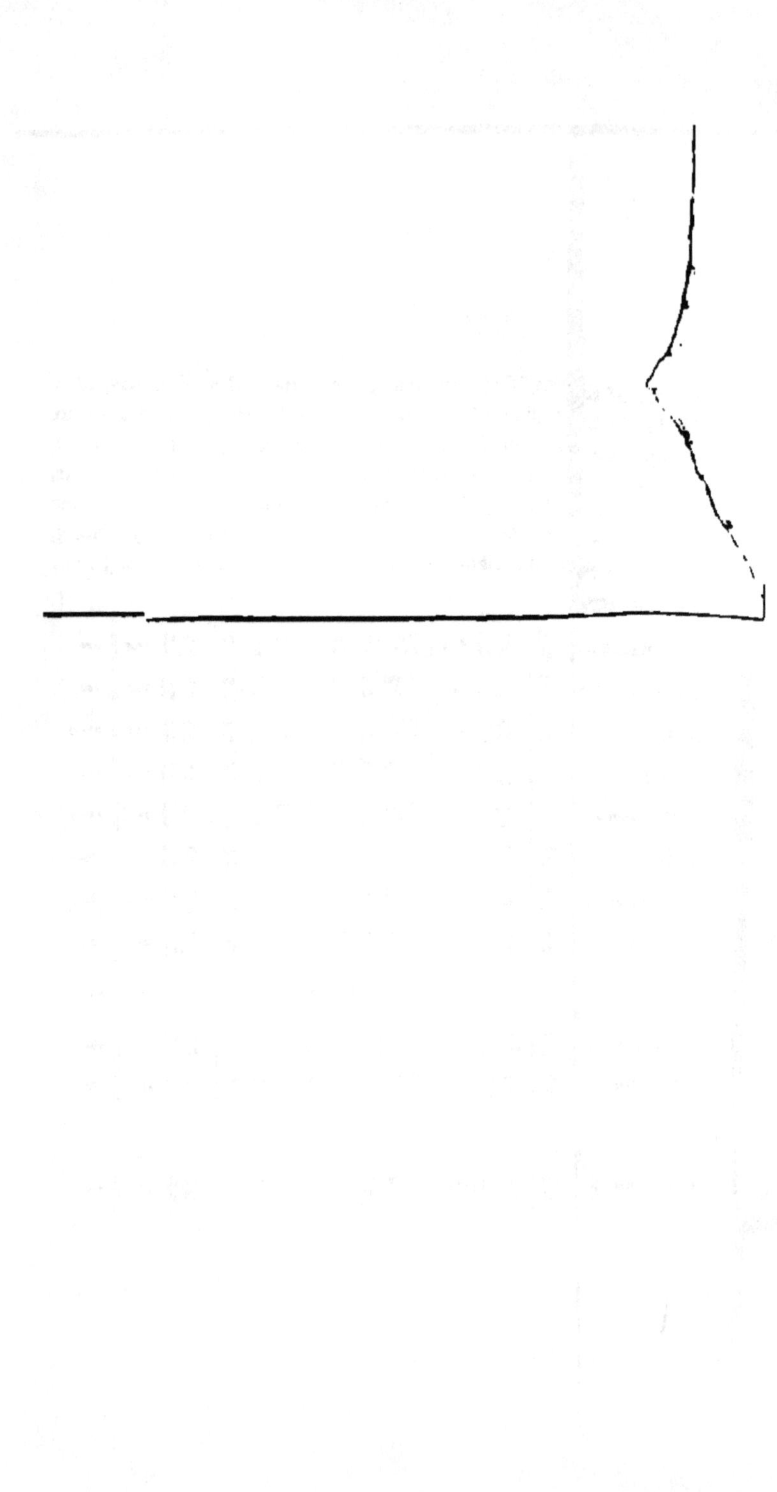

in a well-made, central-fire breech-loader, either snap or double-grip.

THE SIDE-LEVER SNAP-ACTION BREECH-LOADER.

The accompanying illustration, Fig. 35, represents the side-lever snap-action breech-loader, an English-made gun, much admired by American sportsmen. The lever is situated by the side of the lock, and is jointed in the trigger-plate. The lever is very conveniently placed. By pressing it downwards with the thumb of the right hand, the barrels are opened for loading. The action is quick and pleasant, and adapted for rapid firing; these weapons are made generally 10 and 12 bore, and used for duck-shooting. The spring in this gun we consider the strongest and best of all the snap-actions made.

The letter L shows the lever detached; B is the bolt, which is worked by the said lever engaging in the slot; S is the spring which is fixed in the body, in front of the trigger-guard; E is the extractor. These limbs work in the gun in about the same positions as represented in the engraving. The exploding pins are worked by spiral springs. These guns are made also with Greener's patent strikers.

I now introduce the report of the "Field" Gun Trial of 1866, which may interest some sportsmen, being the last public trial of the shooting of breech-loading shot-guns. (*See appended Table.*)

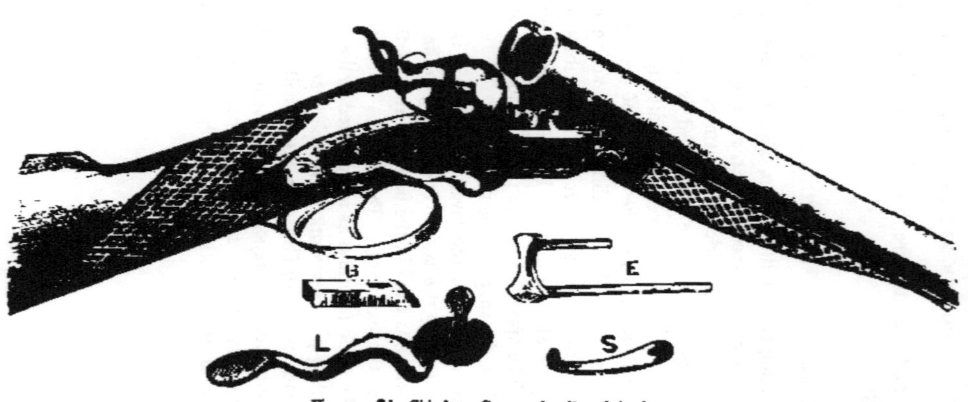

Fig. 35.—The Side-lever Snap-action Breech-loader.

The first seven or eight guns on the list are considered first-class shooting guns, and some pains must have been taken by the makers to get them to perform so well. All the others are considered indifferent; they could not have been properly regulated. The secret lies in knowing how to bore, to make them shoot well. It is not a matter of chance, as some suppose. The shooting of these thirty-two guns may be considered a fair average of all the guns now made and sold by respectable makers both in London and the country. There is only about one-third that shoot really well, the others being second and third rate.

Many sportsmen think if they can kill partridges at twenty-five yards, that they have a good shooting gun; but to be a first-rate gun, it must be capable of killing grouse and pheasants at fifty yards—this is the real test.

Considerable improvement in the penetration of guns has been made since the trial of 1866. We now get the pellets through 30 to 35 sheets of paper. That is an improvement of about 25 per cent. We can also get a better pattern, but we consider this of less consequence than the penetration. A gun that will put 120 shots in a 30-inch target at forty yards and penetrate thirty-five sheets, is to be preferred to one that will put 170 and only penetrate twenty-five sheets. The former will kill game clean and well at a long distance, and will not riddle it at twenty-five yards so as to spoil it for the table.

We give a list of the shooting of some of our best guns from our Shooting Diary of 1869—70, with the numbers of the guns that have been supplied to sportsmen at home and in India:—

SHOT AT A 30-INCH TARGET AT FORTY YARDS, WITH 3 DRAMS LAWRENCE'S NO. 3 GRAIN POWDER AND 1⅛ OZ. SHOT NO. 6.

No. of Gun.	Pattern of Right Barrel.	Pattern of Left Barrel.	Penetration.
9,265	150	130	30
9,267	140	130	30
9,235	130	128	30
9,321	133	151	30
9,281	140	130	30
9,029	130	130	30
8,941	149	160	30
8,968	131	139	30
8,902	144	130	29
8,954	140	151	39
8,959	150	157	25
9,017	180	156	30

The last-mentioned gun we consider an extraordinarily good one. We could give a long list, but the above will be quite sufficient as an example of first-class shooting. We introduce the representation of a target thirty inches centre and four feet square, with a diagram of the shooting of a good gun at forty yards with a charge of 3 drams of powder and 1⅛ oz. of No. 6 shot. There are 146 pellets in the 30-inch circle, 114 on the outside, making 260 on the whole target, and about 30 just over the edge. The remainder of the charge would not reach the target. This is, however, nearly the whole charge of shot in the four feet square. There are about 310 pellets in the charge; the quantity of shot lost in a charge from a good shooting gun is very small indeed. It has been asserted by some, who are opposed to the breech-loading system, that the loss is over 30 per cent., which estimate we have proved to be incorrect. We have counted the shot before loading, and after they have struck the target. The penetration of the gun that made the annexed diagram was 33 sheets of thick Imperial brown paper, the same as that used at the Field-gun Trial. Of course the centre shots are the strongest; the more deviation there is from the centre, the weaker the penetration

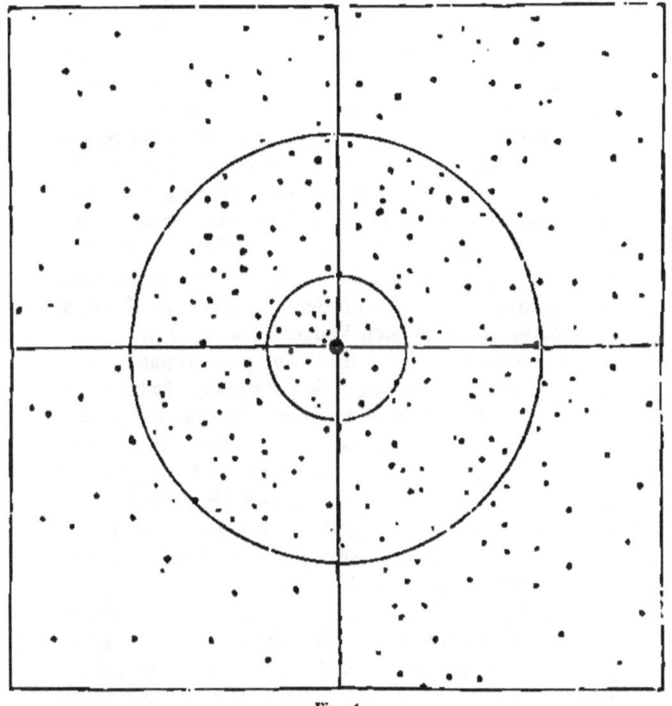

Fig. 36.

becomes. The brown paper test is the best for ascertaining the penetration of a gun. The tin canister test is uncertain and unsatisfactory, as canisters vary so much in thickness and tenacity of material.

Some shot is harder and penetrates the canisters better than the softer kind. We recommend gentlemen to avoid the tin canister and bottle test. If they wish to give a gun a fair trial, let it be at about thirty-five sheets of brown paper, tied at each corner and fastened against a wall or iron target; measure the distance with a tape, because a yard or two, either under or over the forty yards, makes a considerable difference in both pattern and penetration.

We find by experiment that a gun, putting in 125 pellets in a 30-inch circle at forty yards and penetrating thirty sheets, will at thirty-five yards put 155 and penetrate thirty-five sheets; at thirty yards put 185 and penetrate forty sheets; at twenty-five yards put 220 in a 30-inch circle and upwards of 80 more just outside; that is, the whole charge, with the exception of two or three pellets. The number of pellets or corns in 1⅛ oz. of shot, Walker's No. 6, is 308. The penetration at twenty-five yards is very great, averaging forty sheets of paper; that is, a gain in penetration of about one sheet per yard, from forty to twenty-five yards' distance. The pattern is increased about 100 pellets in fifteen yards, which is over six pellets per yard, short of forty. Beyond forty yards both penetration and pattern begin to diminish perceptibly. At forty-five yards we get about 102 pellets and penetrate twenty-six sheets; at fifty yards, 80 pellets and penetrate twenty-three sheets.

In trying experiments with large-size shot at long ranges, we find a great advantage in light 12-bore guns, using less powder and a larger charge of shot; for instance, with No. 2 shot 1½ oz. and 2½ drams of powder shoot well at fifty yards, we get 68 pellets and penetrate twenty-six sheets. No. 3 shot (same charge) is about equal. With No. 4 shot we can get 98 pellets and twenty-five sheets' penetration. For long ranges we recommend 2¾ drams of powder and about 1⅜ to 1½ oz. of No. 4 shot and a concentrator. This will kill game at sixty yards clean. No. 2 and 3 shot, in about the same proportion, will be found equally effective at ducks at about seventy or eighty yards.

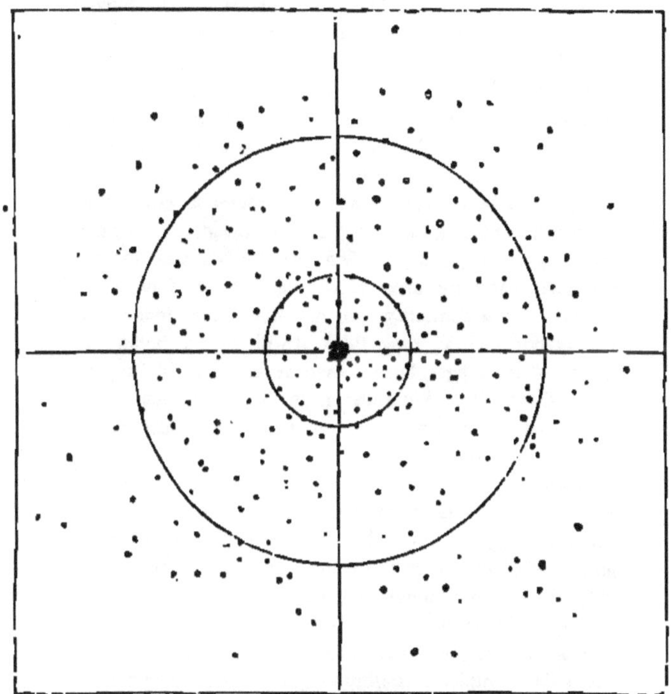

Fig. 37.—A Target at Twenty-five Yards.

THE BORE AND LENGTH OF BARRELS BEST ADAPTED FOR SHOT-GUNS.

Some years ago, in the muzzle-loading days, barrels were used from thirty-two to thirty-six inches long; the general impression being that long barrels killed at a greater distance than short ones. This theory has been exploded; barrels have been shortened through accident, and have proved to shoot better than before reduction in length, much to the surprise of the owners. It does not follow that all guns should be of a short length to shoot well, nor would one length be suitable for all bores; there are so many things that influence the shooting of a gun. Sportsmen should have the length of the barrels made to suit the charge of powder and shot intended to be used.

For a 12-bore breech-loader the general length is thirty inches, for ordinary shooting; but we consider twenty-eight inches quite long enough. Some of the best shooting guns we have made are only twenty-eight inches long in the barrels. Some gentlemen will not have them longer than twenty-four inches; they think it will answer better than having them longer, but we have never been able to prove that satisfactorily. We think it unwise to shorten a barrel more than to twenty-six inches, especially if the gun be required for long distances.

Good shooting guns of 12-bore and thirty inches long we have shortened to twenty-six inches, without injury to the shooting; we have tried them at forty and fifty yards, and satisfied ourselves of the fact. By shortening as above it lightens the gun about six ounces.

We prefer 30-inch barrels for a 16-bore gun. As the bores increase in size the barrels may be shortened in due proportion; for example, a 10-bore weighing 8 lbs., using a charge of 3½ drams and 1½ oz. shot, will shoot equally as well if made twenty-four inches, as if it were thirty inches. The powder would all be consumed in the length.

Long barrels are only desirable for duck-guns, where large charges of powder are used. The weight of gun should be increased in proportion, say ½ lb. to every one inch in the barrel, commencing at 8 lbs. and 24-inch barrel. It is folly to make a 32-inch barrel, 10-bore, to weigh only 8 lbs.; this length and bore should be 11½ to 12 lbs. This would allow of 5 drams powder and 2 to 2¼ oz. shot without inconvenient recoil, and would admit of the action of the gun being made strong at the breech to resist heavy charges. It is unnecessary for sportsmen to encumber themselves with heavy guns unless they want to use heavy charges. Barrels longer than those specified above will be only a detriment instead of an advantage. Large-bore guns, say any size over 12, can be used to advantage only with large charges of shot, 1⅛ oz. and upwards. The larger the bore, the more it scatters.

It is of no advantage to put a large charge of shot in a small bore, as it only increases the column of shot in depth and not in width. The result of this is that the shot gets jammed, and great friction is caused in passing out of the barrel. The proper way is to increase the width with the depth.

Eight-bore guns for duck-shooting should not be made lighter than 12 lbs., and thirty inches long in the barrels would be suitable for the above-mentioned weight.

For breech-loading duck-guns it is important that the barrels should be as short as possible, compatible with good shooting. The longer the barrels are, the greater the strain upon the action, when made upon the Lefaucheux plan. The hinge-pin upon which the barrels turn must be made near to the breech, about 2½ to 3½ inches. Every inch added to the muzzle increases the strain upon the breech action, and makes the gun more liable to get wide at the breech and to droop at the muzzle. The levers also work much stiffer, owing to the length of barrels. There is a remedy for this by having a longer body, and getting the hinge joint further away from the breech, according to the weight and size of bore. This is more trouble and expense to the gun-maker, and increases the cost of the gun; but it is the proper way to have these guns made, and the only way to ensure durability.

The above plan is essential in all systems, whether levers or snap actions. We have more faith in this than in the so-called treble grip, so highly spoken of in some of the sporting papers.

THE BEST MATERIAL FOR GUN-BARRELS.

Many years ago, W. Greener brought out the laminated steel as the very best and most suitable metal for gun-barrels. At first many gun-makers ridiculed the idea of using such a hard metal, but it is most remarkable how many have come round to the same opinion, and the laminated steel has now a world-wide reputation. Those makers who do not profess to make guns of that material use a stub damascus, which is only an inferior quality of laminated steel. As it is now made, the damascus barrels that are so much used by the London makers contain considerably more steel than those made years ago.

It is an established fact that hard barrels can be made much lighter, that better shooting can be got out of them, that they are more lasting, and that they retain their superior shooting qualities longer than those made from soft metal.

To enable the reader to form an accurate opinion of the steel used for gun-barrels, it will be necessary for me to describe the whole process from the beginning. Our laminated steel is made in the following manner:—

Having collected a sufficiency of mild steel scraps, such as cuttings of saws, waste from steel pen making, old coach springs, and the immense variety of pieces arising from the various manufactures of tools; they are cut into pieces of equal dimensions, polished in a revolving drum by their friction against each other, until quite bright, and then placed for fusion on the bed of an air furnace. The parts first fused are gathered on the end of a similarly fabricated rod, in a welding state, and these gather together, by their adhesion, the remainder, as they become sufficiently heated, until the bloom is complete. The steel is then removed from the furnace, and undergoes the effect of a three-ton forge hammer and the tilt, until it forms a large square bar; it is then re-heated, and conveyed to the rolling mill, where eventually it is reduced to the size of rod desired. We generally have the metal required cut into pieces of six inches in length. A certain number are bundled together, welded, and then drawn down again in the rolling mill. This can be repeated any number of times—elongating the fibres and multiplying their number to an indefinite extent, as may be required.

The great advantage derived in this instance from air-furnace welding is a chemical one; for, while the small pieces of steel are fusing on the bed of the air furnace, the oxygen is extracting the carbon, and leaves the resulting metal mild steel, or iron of the densest description, while the succeeding hammering, rolling, and re-welding cause the formation of a mass of metallic fibres. The polishing secures a clear metal; indeed, so free from specks are the generality of barrels thus made, that it is scarcely possible to imagine clearer metal. When contrasted with the best of ordinary iron by the aid of a powerful microscope, its superior closeness and density of grain are markedly apparent.

To such an extent has this process been carried, that we can produce specimens of a considerably increased specific gravity. The barrels made of this metal, in general, surpass all tried against them, with this great advantage, that the finer the polish in the interior the better they shoot, and the longer will they remain free from lead. The only difficulty is in the working, as the boring, filing, &c., are more difficult. Moreover, greater care is required to see that they are not annealed when in the hands of the borer or filer; for in such case they would be considerably injured, though not to the same extent as barrels of a softer nature. We tested a great variety of bars by drawing them asunder longitudinally with the testing machine; and the average strength of a rod of $\frac{6}{16}$ broad by $\frac{5}{16}$ thick and 12 inches long, containing 1·40625 inch of solid iron, was equal to a tension of 11,200 lbs. This furnished a barrel having a thickness of metal in all parts of the arch equal, or $\frac{3}{16}$ of an inch thick, capable of bearing an internal pressure of 6,022 lbs. to the inch of the tube.

The generality of barrel-makers spoil this metal by an attempt to obtain figure; for all extreme twistings in the rod depreciate the metal by separating the fibres: to borrow a simile, they obtain only an over-twisted rope. This is not only disadvantageous, but useless; for the extreme density of the metal renders the figure difficult to be shown distinctly, as acid acts upon it but slightly, and never so well as on metal fabricated from two differently constructed carbonised materials.

The laminated steel now so universally made and used by gun-makers, is manufactured by a different process to that I have just described, and that which we will call the modern plan.

The gun-iron makers now get a long strip of mild steel the thickness required, and then a bar of superior iron the same size; then another of steel, and so on to the number of twelve, laying them upon each other alternately. The whole are then welded together, and drawn through rolls, which reduces the size to $\frac{3}{8}$ or $\frac{5}{8}$, as required, and into square bars. In this form it is supplied to the barrel-welders, who work it up into the barrel. This process we will proceed to describe.

When about to be converted into damascus, the rod is heated the whole length, and the two square ends put into the heads (one of which is a fixture) of a kind of lathe, which is worked by a handle similar to that of a winch. It is then twisted like a rope, until it has from twelve to fourteen complete turns in the inch. By this severe twisting, the rod of six feet is shortened to three, doubled in thickness, and made perfectly round. Two of these rods are placed together, with the inclinations of the twist running in opposite directions. They are then welded into one, and rolled down into a rod $1\frac{1}{8}$ of an inch in breadth. This rod is twisted into a spiral tube by attaching one end to a mandril, which is turned round by a handle until the whole strip is coiled. It is usual to make a gun-barrel in two parts. The breech is made of thicker strips than the muzzle.

They now begin the welding. Spirals that are intended for the breech-end are heated to a welding-heat for about three inches, removed from the fire, and jumped close by striking the end against the anvil. Again they are heated and jumped, to insure perfect welding. They are then beaten lightly in a groove to make them round.

The neatest part of the process consists in joining the points of the two rods, so as to make the barrel appear as if it had been twisted out of one rod. The ends of the two rods are a little detached, brought from the fire, and applied to each other; a gentle tap is then given, and the union is perfect in an instant. The rapidity and dexterity with which this is accomplished ought to be seen, to be duly appreciated. This trouble is only taken with the best barrels. In the manufacture of barrels of an inferior description, the ends of the rods are cut in a sloping direction, and when welded together, become quite square at the part where the pieces are joined. In a finished barrel the points of junction are easily recognised. By tracing the twist, a confusion will be found to exist for about an eighth of an inch, every six or seven inches; and from this appearance you may conclude that, for a barrel so joined, the welder had not the best price. Having joined the whole of the spirals, three inches are again heated to

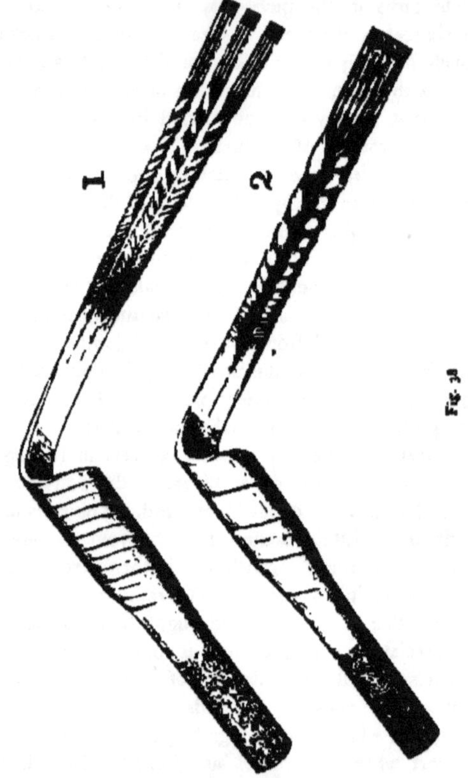

a welding heat, the mandril is introduced, and the tube hammered, in a groove, to the size required. This operation is repeated until the whole length is finished.

Then follows hammer-hardening—that is, beating the barrel, in a comparatively cold state, in a groove, with light hammers, for the space of half an hour. This is a most important part of the process. It closes the pores, condenses the texture of the metal, compresses a greater substance into less bounds, increases greatly the strength of the barrel, and renders it more elastic.

Some persons have an idea that laminated steel barrels are often spoiled in the course of welding by being burnt in the fire during the process; but should this be the case by negligence, it would be easily discovered, and the barrel so injured would not reach the next stage of manufacture, so glaring would be the roughness of the metal from its being honey-combed.

English damascus is made much in the same way as the laminated steel just described. It contains nearly as much steel, and is almost as hard; but the strips or bars of iron and steel are arranged so that the figure may be more clearly defined, as represented in the accompanying sketch. The white marks are the steel, and the dark ones the iron. The curly figure is obtained by twisting the rods as before described. Damascus has three rods, which form the strip to make the barrels; this causes the figure to be finer. Laminated steel has but two rods. Many gun-makers prefer damascus merely because they think that metal is clearer from specks or greys; but we find no difference in that respect, in metals made as we have described.

Belgian damascus barrels are now used extensively in England by nearly all the gun-makers. We consider them much inferior to the English make, being too soft for really good gun-barrels. They have not the resisting power, and do not shoot so well or last so long as the English damascus. They may easily be detected; the figure is much finer and less distinct, in consequence of there being a larger proportion of iron in the mixture. The fine figure is caused by the rods being smaller before being twisted. It is usual to make barrels of three rods; sometimes four and five are used, making a variety of fancy figures.

GUNPOWDER.

The nature of black gunpowder is so well known in the sporting world, that we think it would be a work of supererogation on our part to attempt to give a description of it, especially after our father's article on the subject, which we quote from his last work on gunnery. Little or no improvement seems to have been made in granulated gunpowder for sporting purposes for many years, and we think we may venture to say that it will remain *in statu quo* for some years to come, and, in spite of all the new explosives, will hold its own.

Gunpowder, whether considered relatively to engines of war, or to those arms used with so much success in the sporting field, has, since its first *introduction*, been a source of much and frequent discussion. In regard to its origin, we shall not much enlarge, nor repeat the many suppositions and conjectures promulgated by the searchers after antiquarian evidence.

The inhabitants of India were unquestionably acquainted with its composition at an early date. Alexander is supposed to have avoided attacking the Oxydracea, a people dwelling between the Hyphasis and Ganges, from a report of their being possessed of supernatural means of defence; "for," it is said, "they come not out to fight those who attack them, but those holy men, beloved by the gods, overthrow their enemies with tempests and thunderbolts shot from their walls;" and, when the Egyptian Hercules and Bacchus overran India, they attacked these people, "but were repulsed with storms of thunderbolts and lightning hurled from above." This is, no doubt, evidence of the use of gunpowder; but as it is unprofitable to investigate this subject further, we shall merely confine ourselves to the European authorities.

Many ascribe the discovery of gunpowder to Roger Bacon, the monk, who was born at Ilchester, in Somersetshire, in the year 1214, and is said to have died in 1285. No doubt he was by far the most illustrious, the best informed, and the most philosophical of all the alchemists. In the 6th chapter of his Epistles of the Secrets of Arts, the following passage occurs:—"For sounds like thunder, and flashes like lightning, may be made in the air, and they may be rendered even more horrible than those of Nature herself. A small quantity of matter, properly manufactured, and not larger than the human thumb, may be made to produce a horrible noise; and this may be done many ways, by which a *city* or an *army* may be destroyed, as was the case when Gideon and his men broke their *pitchers* and exhibited their lamps, fire issuing out of them with great force and noise, destroying an infinite number of the army of the *Midianites*." And in the 11th chapter of the same epistle occurs the following passage:—"Mix together saltpetre with *luru mòne cap ubre*, and sulphur, and you will make thunder and lightning, if you know the method of mixing them." Here all the ingredients of gunpowder are mentioned, except charcoal, which is, doubtless, concealed under the barbarous terms used; indeed, the *anagram* is easily converted into *carbonum pulvere*, with a little attention.

This discovery has also been attributed to Schwartz, a German monk, and the date of 1320 annexed to it—a date posterior to that which may be justly claimed for Friar Bacon; and an accident is stated to have been the means by which he discovered it.

Mr. Hallam, referring to the authority of an Arabic

author, infers that there is no question that the knowledge of gunpowder was introduced into Europe through the means of the Saracens, before the middle of the thirteenth century; and no doubt its use then was more for fireworks than as an artilierist projectile force. There is good evidence, too, that the use of gunpowder was introduced into Spain by the Moors. The Chinese are also reported to have been acquainted with a similar explosive at a very early period.

The composition of gunpowder, as regards the proportions of the ingredients, has not undergone any material alteration, the chemical proportions of the ancients being nearly those of the present day.

Gunpowder is an explosive propellent compound. The terms *explosive* and *propellent* are not here used as synonymous—they are not convertible; for a chemical mixture may possess the *explosive* power in a much higher degree than the *propellent*. Fulminating gold, silver, and mercury are dreadfully explosive; but they have not the same projectile force, nor can they be used as a substitute for it. Several experiments have been made with compounds of this nature, but the result is the reverse of what might be expected. Nothing can resist the exceeding intensity of the action of fulminating powder. A shot, when fired in this way, is not projected as by gunpowder, but is split into fragments by the velocity of its explosion, as we shall hereafter have occasion to show.

Nitre, or saltpetre, is strictly the essence of gunpowder. It is a triple compound of oxygen, nitrogen, and potassium. The chemical action of those elements on each other, and the play of affinities between them at a high temperature, occasion the immense effect produced by gunpowder on the application of fire or heat. By universal consent, sulphur is included in the mixture; but it is not absolutely necessary for the "propellent power," for nitre and charcoal only will generate effects similar to the compound with sulphur. Gunpowder made without sulphur has, however, several bad qualities. It is not on the whole so powerful, nor so regular in its action; it is also porous and friable, possessing neither firmness nor solidity. It cannot bear the friction of carriage, and in transport crumbles into dust. The use of sulphur, therefore, appears to be not only to complete the mechanical combination of the other ingredients, but being a perfectly combustible substance, it increases the general effect, augments the propellent power, and is thought to render the powder less susceptible of injury from atmospheric influence.

Gunpowder consists of a very intimate mixture of sulphur, carbon (charcoal), and nitrate of potash (nitre). The proportions in which they exist are two equivalents of nitre, one of sulphur, and three of carbon (new notation). The great explosive power of gunpowder is due to the sudden development from its solid constituents of a large quantity of gases; these gases are nitrogen and carbonic acid.

TABLE OF COMPOSITION OF DIFFERENT GUNPOWDERS.

Mills.	Nitre.	Charcoal.	Sulphur.
Royal Waltham Abbey . . .	75·00	15·00	10·00
France, National Mills . . .	75·00	12·50	12·50
French Sporting	78·00	12·00	10·00
French Mining	65·00	15·00	20·00
U. S. of America	75·00	12·50	12·50
Prussia	75·00	13·50	11·50
Russia	73·78	13·59	12·63
Austria (Musket)	72·00	17·00	16·00
Spain	76·47	10·78	12·75
Sweden	76·00	15·00	9·00
Switzerland (Round Powder) .	76·00	14·00	10·00
Chinese	75·00	14·40	9·90
Theoretical proportions as above	75·00	13·23	11·77

At the ordinary temperature of the atmosphere these gases would occupy a space 300 times greater than the bulk of the gunpowder used; but, owing to the intense heat developed at the moment of explosion, they occupy at least 1,500 times the bulk of the original gunpowder. The mixture, consisting of two equivalents of nitre, one of sulphur, and three of carbon, would yield three equivalents of carbonic acid, two of nitrogen, and one of sulphide of potassium. The change may be represented thus—

$$S+C_3+2KNO_3=3CO_2+N_2+K_2S.$$

The only solid residue, therefore, is the sulphide of potassium, and this is the compound which produces the sulphurous odour on washing out a gun-barrel; water is decomposed, sulphuretted hydrogen and potash being the result of the decomposition.

Gunpowder is now made by all the sporting gunpowder manufacturers from No. 1 to No. 5 grain; and it appears certain that a further increase in the size of the grain would be advantageous; for many years of patient and laborious experiment clearly show, that the old notion of gunpowder being blown out of an ordinary-sized gun in an unburnt state is one of the "purest of vulgar errors:" such a thing, indeed, cannot possibly happen unless the powder be bad, or the gun *imperfectly made* or injudiciously charged.

I am satisfied that I am under rather than over estimate, when I assert that six drams of ordinary sporting gunpowder may be completely exploded in a barrel of 14 bore, 2 feet 6 inches long, with a resisting projectile one ounce in weight above it. This, however, being more than a double charge for such a gun, cannot be pleasantly practised; and it is only asserted by way of argument.

Assuming then, for argument's sake, that six drams of gunpowder are exactly consumed in passing from the breech to the muzzle of a gun 2 feet 6 inches long, and that the shot therefore acquires its greatest velocity as it leaves the muzzle, it follows that the ordinary charge of 2½ drams will be wholly consumed before it has traversed half the length of the barrel, and consequently the charge of shot must here acquire its greatest velocity. It is certain, then, that the shot must travel the latter half of the barrel at a diminished velocity, and its velocity must continue to diminish as it passes up the barrel, for two obvious reasons—1st, the column of air in front of the charge is more condensed, and thus offers a greater resistance to the exit of the charge; 2nd, the velocity is continually diminished by the increased friction of the charge against the barrel.

The perfection of projectile science is to make the projectile acquire its greatest velocity at the instant of leaving the muzzle; and if, by increasing the size of the grain of gunpowder, we can diminish the rapidity of its explosion—thus causing it to burn and generate fresh gas up to the muzzle of the gun—the projectile will then acquire its greatest velocity, and leave the gun to the best advantage: this is the important point which has hitherto been overlooked, not only in fowling-pieces, but in the expansive principle of rifles.

In order to obtain the best results from a gun, the gun itself must be perfect in construction, and the expellent force must be brought to bear in the best possible manner upon the projectile; and this is to be done by attending to the granulation of the powder, which must be suited to the length of the gun, to its bore, and to the weight of the projectile.

Common-sense, and engineering skill, will demonstrate that according to the weight of matter to be projected must be the nature of the expellent: *accumulative*, until it has overcome the inertia of that matter; *acceleratire*, until it has communicated to it the highest state of velocity its power is capable of effecting. If, on the other hand, it is inferior to this, science has not extracted from it the full *horse-power* it contains; and we are uselessly expending force and destroying our engines by undue pressure being exerted on one part, and inferior pressure on another; whilst by a proper distribution of that force durability of the cannon is insured, and from twenty-five to thirty per cent. more work may be obtained from an equal quantity of powder, provided its granulation be judiciously selected according to the area of the gun.

Another advantage of using gunpowder of a suitable granulation is the absence of sharp recoil; and thus greater accuracy of range is obtained—accuracy of range and steadiness of weapon being inseparable.

Large-grain gunpowder is not only a more effectual expellent than the fine-grain, but is much more safe to use, for by using it the risk of bursting the barrel is greatly lessened, as a very simple illustration will show. If we estimate the force generated by the usual charge of 2¾ drams (I confine the question to the 14-bore gun, for the sake of uniformity) to be 5,000 lbs., whether the powder be fine or coarse grain, it follows that the fine powder, igniting so rapidly, will exert all its force on the breech-end of the gun; whereas the coarse powder, igniting less rapidly, distributes this force over the whole length of the barrel: hence the greater risk of a gun bursting with fine powder than with coarse. If we suppose the fine powder to be entirely ignited when it reaches half-way up the barrel, then the force of 5,000 lbs. is exerted on the lower half of the barrel; but if the coarser grain is not entirely ignited until it reaches the muzzle, then the force of 5,000 lbs. will be distributed over the whole length of the gun.

But this is not all. The fine powder, igniting almost instantaneously, exerts its force in all directions at once, and the barrel may burst at the side before the charge has time to move: whereas the coarse powder, igniting as it does more slowly, first lifts the charge, and then the volume of gas behind it, increasing as the powder becomes more thoroughly ignited, sweeps the charge out of the barrel with a velocity increasing towards the muzzle.

If time is not given for the charge to receive the full advantage of the expansive force of the generated air, the force is exerted not upon the charge, but upon the barrel of the gun itself; and that time is necessary for the full development of this force, is proved by the fact that miners mix their gunpowder with sawdust, in order to diminish the rapidity of its explosion, and thus get the advantage of its force in the distance; from the miners, then, let us learn how to obtain the greatest benefit from this force, and waste it not.

There can be no doubt of the importance of this principle; little progress has, however, been effected, from want of scientific illustration. Let it be defined like that of steam power, and its adoption will follow as a natural consequence.

For several years I have had gunpowder manufactured of various sizes, at the sight of which most sportsmen would express their astonishment.

One objection held by sportsmen to the large-grained gunpowder is that it does not come up to the nipple of the gun; now, although I do not consider this at all important, still if the specific gravity of the gunpowder were increased by compressing 1½, 2, or 3 grains of gunpowder into the space of 1 grain, by means of hydraulic pressure, this objection would at once be set aside; whilst at the same time the powder would be less liable to absorb moisture, or to become friable with age: either of which conditions is incompatible with good shooting.

The granulating of gunpowder, to be most perfect, should be on a uniform principle; the manipulation should be alike in all particulars, but especially in that part of the process which determines the specific gravity. The hydraulic pressure on the cake should be alike in all cases: in fact, the various sizes of grain might be produced from the same cake, and the desired object be thus obtained. But so long as the practice is followed of producing large grain from less condensed cake, the article produced will give unsatisfactory results; and the advantages which might be attained, as our experience denotes, and which would be of the greatest service, alike in sporting rifle and artillery powder, will be nullified.

There is an instrument used by some sportsmen, and strongly recommended by many gun-makers, for testing the strength of different kinds of gunpowder. It consists of a chamber closed by a spring, and fired like an ordinary pistol. When the powder explodes, the spring is forced forward, and moves an index round a graduated circle; the more quickly the powder explodes, the farther does it lift the spring; hence this is a measure of quickness of fire, but not of expellent force; and from the observations which have been made on gunpowder, it must be evident to any one who has paid the least attention to the subject, that this instrument is utterly useless.

An instrument to test the comparative strength of different kinds of gunpowder is yet a desideratum in projectile science; and we cannot doubt that such an instrument will be produced, when the importance of the granulation of gunpowder is more generally known and appreciated.

THE MANUFACTURE OF GUNPOWDER.

The charcoal formerly used was made in the common way by pits, which must have been seen by almost every one. The method is now to *distil* the wood in cast-iron cylinders, extracting the pyroligneous acid, &c., by heating them red-hot, and allowing all other volatile matter to evaporate, the charcoal only being retained in the cylinders or retorts; hence arises the name of *cylinder gunpowder*. The best charcoal for sporting powders is the black dogwood; Government use willow and alder; and ordinary charcoal does for common powders. Charcoal is ground in the same way as the nitre. Sulphur is purified simply by fusing, and when in that state skimming off the impurities; it is cooled and pulverised in the same way as the other two ingredients. The three ingredients, after being carefully weighed in their due proportions, are sifted into a large trough, and well mixed together by the hands. They are then conveyed to the powder-mill. This is a large circular trough, having a smooth iron bed, in which two millstones, secured to a horizontal axis, revolve, traversing each other, and making nine or ten revolutions in a minute. The powder is mixed with a small quantity of water put on the bed of the mill, and there kept subject to the pressure of the stones; and if we calculate the weight of the two millstones at six tons, it follows that in four or five hours' incorporation on this bed, it subjects the ingredients to the action of full 10,000 tons. It is this long-continued grinding, compounding, and blending together of the mixture, that renders it useful and good. After this intimate mixing, it is conveyed away in the shape of mill-cake, and firmly pressed between plates of copper. Bramah's press has been introduced of late years—we should say with a great deal of improvement to the powder, as will be shown hereafter—and by its means the mass is more compressed and in thinner cakes. It is then broken into small pieces with wooden mallets, and taken to the corning-house, where it is granulated, "by putting it into sieves, the bottoms of which are made of bullocks' hides, prepared like parchment, and perforated with holes about two-tenths of an inch in diameter; from twenty to thirty of these sieves are secured to a large frame, moving on an *excentric* axis, or crank, of six inches throw; two pieces of lignum vitæ, six inches in diameter, and two inches or more in thickness, are placed on the broken *press-cakes* in each sieve. The machinery being then put in rapid motion, the discs of lignum vitæ (called balls) pressing upon the powder, and striking against the sides of the sieves, force it through the apertures, in grains of various sizes, on to the floor, from whence it is removed, and again sifted through finer sieves of wire, to separate the dust and classify the grain. One man works two sieves at a time, by turning a handle and excentric crank; the sieves being fixed to a frame, which is suspended over a bin by four ropes from the ceiling."

The grains afterwards undergo a process of *glazing*, by friction against each other, in barrels containing nearly 100 lbs., making forty revolutions in a minute, and lasting several hours, according to the fancy of the purchaser. This part of the business we entirely disapprove of, as injurious to the quick and *certain ignition*. Gunpowder is finally dried by an artificial temperature of 140° Fahrenheit, which is suffered gradually to decline. The last process is sifting it clear of dust, and then packing it in canisters or otherwise.

Fine grain powder, when unconfined, explodes more quickly than large, or is sooner burnt out, and consequently generates more force in the same period of time; but when it comes to large quantities, its very quickness is detrimental to its force, by condensing the air around the exterior of the mass of fluid, which thus constrains its bound. In small quantities, the proportion of condensation is not so apparent, and hence the reason why greater velocities can be obtained with small arms than with cannon.

There exists a diversity of opinion in regard to the strength or projectile force of gunpowder. Dr. Ure remarks—"If we inquire how the maximum gaseous volume is to be produced from the chemical reaction of the elements of nitre on charcoal and sulphur, we shall find it to be by the generation of carbonic oxide and sulphurous acid, with the disengagement of nitrogen. . . It is obvious that the more sulphur, the more sulphurous acid will be generated, and the less forcibly explosive will be the gunpowder. This was confirmed by the experiments at Essonne, where the gunpowder that contained twelve of sulphur, twelve of charcoal, in 100 parts, did not throw the proof-shell so far as that which contained only nine of sulphur and fifteen of charcoal. The conservative property is, however, of so much importance for humid climates and our remote colonies, that it justifies a slight sacrifice of strength."

"When in a state of explosion, the volume," Dr. Hutton calculates, "is at least increased eight times, and hence its immense power. The pressure exerted, if in a state of confinement, will depend on the dimensions of the vessel containing it; so that it would be no difficult undertaking to obtain any pressure above that of the atmosphere, up, we may fearlessly say, to the enormous amount of 4,000 lbs. per square inch."

The same quantity of gunpowder, subjected to a variety of experimental tests, differs materially in its results; at the same time it is only by such a method that we can arrive at the relative strength or power which it possesses. Dr. Hutton, whose authority in all mathematical calculations is very high, and whose opinions and judgment in matters of this nature ought not to be unthinkingly controverted, states 2,000 feet per second (with cannon) as the highest velocity which any projectile had attained, at the time of his writing, that had gunpowder for its propellent power.

In all uses of gunpowder, the grain should be of a size proportioned to the length and bore of the gun; for if we have not an accelerating force to overcome the increasing resistance of the compressed column of air in the barrel, there is great danger that the gun may be burst, and probably be productive of great mischief; whilst a judicious application of the extraordinary power thus placed at our disposal, may be alike conducive to our safety and our pleasure. A musket-ball can be driven through a half-inch boiler plate; but this can only be accomplished by using as much powder as will generate a gradually, though rapidly, increasing power, until the ball has passed the limits of the tube.

Nitre is not the only salt which has been employed in the manufacture of gunpowder. Its quantity or proportion in the mixture has been lessened, and the deficiency supplied by another elementary combination, namely, by the chlorate of potassa.

It is obvious, from the extremely high character English sporting gunpowder has obtained all over the world, that considerable improvement must have been effected by the private manufacturers, either in the purification or manipulation of ingredients; indeed the unwearied care bestowed on this point by several of our best makers is beyond all praise.

"Granulation," properly understood, is an equivalent point to either chemical or mechanical knowledge and manipulation in gunpowder manufacture. Great anxiety to meet the wishes of the sporting world on this point, and to advance with the age, has been aroused; and specimens have been kindly furnished to us, not by one, but by all the following celebrated makers:—Messrs. Pigou and Wilks, Curtis and Harvey, Lawrence and Son, John Hall and Son; and I have received also a very excellent specimen from the Scotch mills.

Gunpowder is made of five sizes of granulation, on the basis before alluded to: namely, No. 2, containing two quantities of No. 1; No. 3, three, and so on in progression; but it is imperative that all the various sizes be produced from the same mill-cake, or be otherwise of the same condensation or specific gravity; and in all experiments of comparison, equal weights are a *sine quâ non*, otherwise the comparison will be futile; as measure is, for these very obvious reasons, inapplicable in comparative tests. When these points are carefully attained, increased power of killing, "decreased recoil," and much greater safety, will be the important benefits which the gunpowder manufacturers will confer on every one using a gun.

GUN-COTTON.

Gun-cotton has been before the world for some years, but, except as a curiosity, it has attracted little public attention; neither has it gained any reputation as a projectile force. It may be prepared by steeping cotton-wool for a few minutes in a mixture of nitric and sulphuric acids, thoroughly washing, and then drying at a very gentle heat. It consists chemically of the essential elements of gunpowder, viz., carbon, nitrogen, and oxygen; but, in addition, it contains another highly elastic gas, hydrogen. The carbon in the fibres of the wool presents to the action of flame a most extended surface in a small space, and the result is an explosion approaching as nearly as possible to the instantaneous: in consequence of its rapid ignition it produces a violent kick. Sufficient time is not given to put heavy bodies in motion, hence it cannot be usefully employed as a projectile agent. No one who values his limbs should trifle with it, for fearful accidents have resulted from its exposure to the heat of the sun, and other very simple causes.

Although we agree substantially with our late father's opinion on gun-cotton, it has been greatly improved upon since 1853, and made better adapted for sporting purposes by the perseverance of Mr. Prentice. The charges are made up now, and enclosed with india-rubber or gold-beater's skin, to protect them from atmospheric change, which alters the strength and makes the cotton uncertain. Unless some means can be found to equalise the force of the explosion, we fear it will have but a limited sale. The makers of this material would do well to look to this matter, if they wish to make it a commercial success for sporting purposes. Being so quick in ignition and combustion, it exerts an extra strain upon the breech-end of the gun, which is dangerous. If by any means they can make it slower in combustion, so as to equalise the strain in a greater length of the barrels, and make it more certain and equal in strength, it would be far better, and no doubt have an extended sale. It has several advantages over the black powder—for instance, absence of smoke, and cleanliness.

The following letter, which appeared in the *Field* newspaper of April 13th, 1867, will explain the nature of gun-cotton. From experiments conducted by Professor Abel at Woolwich Arsenal, a paper was read by him on this subject at the meeting of the Royal Society on April 4th, 1867:—

Mr. Abel, after giving an abstract of the very various and conflicting statements made by different observers as to the stability of pyroxiline under various conditions of temperature, exposure to light, moisture, &c., went on to detail the results of the experiments conducted under his supervision at Woolwich Arsenal. His conclusions may be briefly stated to be these: Gun-cotton, when carefully prepared according to the minute directions of Baron Lenk, but with the powers of silicating omitted, is stable when kept in the dry state in the dark, and still more stable, if that be possible, when kept wet. The same material, when exposed to sunshine, even for many months, undergoes no material change; but the alteration is somewhat greater if the cotton be damp than if it be dry. When heated to the temperature of boiling water in closed vessels, red fumes are given off after a time, which varies with different specimens; this evolution of nitric peroxide is, of course, an indication of the breaking-up of the compound—a breaking-up which sometimes becomes so rapid as to end in explosion. As was said, some specimens resisted the action of heat for a much shorter period than others, and this was found to be due to the fact that the cotton employed in their manufacture had not been so well freed from gummy and resinous matters in these instances. The products of the action of the mixture of nitric and sulphuric on these resins, when the cotton is immersed, are bodies which are much more unstable than the product of the action of the acids on the true woody fibre; and not only are they themselves prone to decomposition, but when decomposing seem to induce the true gun-cotton to decompose also. Now, it was found that even prolonged boiling of the original cotton fibre with soda failed to completely dissolve out the resins, &c.—a fact which is due to the structure of the fibres of cotton, which, as is well known, are hollow tubes, so far choked or closed that liquids cannot well pass in or out of them. Tearing the cotton up into very short lengths—making a pulp of it, in fact—gets over this difficulty; at any rate, to such a degree that gun-cotton made of the purified pulp is practically unalterable by any amount of exposure to variations of temperature, within moderate limits, and also by the action of light. An additional safeguard against alteration is to be obtained by making the gun-cotton very slightly alkaline, by dipping it in a weak solution of carbonate of soda before finally drying it. The action of this last is to neutralise any trace of acid which may be given off at first by the decomposition of remaining impurities, and so to prevent the peculiarly pernicious effects of the presence of free acid.

Professor Abel concluded by expressing an opinion that no conditions which can obtain in actual service would produce the least injurious effect on gun-cotton, while in many respects this material presents advantages of no mean kind. One of these advantages he illustrated in a rather startling manner. Taking from a large tin box as much gun-cotton as his hand would hold, he remarked that it was slightly damp—not wet, be it observed, but damp only; then, still holding it, he pressed on to its surface a red-hot poker. That it did not explode, my writing of this letter is sufficient evidence. C. H. G.

Many letters appearing in the *Field* newspaper, from sportsmen asking for some reliable information respecting gun-cotton and its adaptability for sporting purposes, we determined to give it a good and fair trial against black powder, and we here give the result as it appeared in the *Field*, September 14th, 1867.

TRIAL OF GUN-COTTON AGAINST GUNPOWDER.

SIR,—For the information of your correspondent "Dotterel," and other gentlemen interested in the subject, I give the result of three trials at my shooting ground of gun-cotton and gunpowder, fired from the same barrel. Charge of powder, 3 drams (Lawrence's No. 3 grain); shot No. 6, 1⅛ oz. Gun-cotton (Prentice's) charge equal to 3 drams powder, and 1⅛ oz. of shot No. 6. Fired at a 30-inch target at 40 yards, with paper for penetration same as used at the "Field" Gun Trial.

	Central-Fire.				Pin Gun.				Central-Fire.			
	Gun-cotton.		Powder.		Gun-cotton.		Powder.		Gun-cotton.		Powder.	
	Patt.	Penet.	Patt.	Penet.	Patt.	Penet.	Patt.	Penet.	Patt.	Penet.	Patt.	Penet.
	136	26 ...	106	31	119	20 ...	160	28	114	20 ...	160	30
	150	28 ...	130	27	110	19 ...	140	32	63	22 ...	136	35
	145	20 ...	123	25	105	14 ...	113	32	43	21 ...	91	29
	127	12 ...	112	30	101	22 ...	106	30	72	19 ...	96	32
	127	25 ...	121	30	50	15 ...	69	25	154	20 ...	100	29
	128	27 ...	143	26					76	14 ...	116	28
Total ...	813	138 ...	735	169	485	90 ...	588	147	522	116 ...	699	173
Average	135	23 ...	122	28	97	18 ...	117	29	87	19 ...	116	30

The general average of the whole is as follows :—

	Pattern.	Penetration.
Gun-cotton	106	20
Powder	118	29

I also tried 2½ drams of powder, and made better penetration than I could get with the gun-cotton. W. W. GREENER.

St. Mary's Works, Birmingham.

We must mention here that when experimenting with gun-cotton against gunpowder, we fired off two barrels with gun-cotton, with little or no penetration, and scarcely any shot reached the target "which we did not record." It will be seen in the above report that one shot with gun-cotton made 150 pattern and fair penetration. This shows the uncertain nature of this explosive; and, as we have pointed out before, gun-cotton must be made more equal in its strength, or regular shooting cannot be made with it.

SCHULTZE'S SAWDUST POWDER AND PYRO-PULVER.

From a series of experiments with these compounds used in our own guns, we consider there is little or no difference in the results obtained.

We think the following will explain the nature of both these compounds, and the only difference we can detect in pyro-pulver and Schultze's Sawdust Powder is that pyro has less specific gravity, runs less freely, and requires half the charge to be rammed into the cartridge-case at a time.

Pyro-pulver is a new form of wood gunpowder. Both explosives are chemical compounds, having wood in small particles for their base, but produced in different ways.

By Schultze's method, wood in grains, being like gun-cotton converted into pyroxyline by the usual treatment with nitric acid, and continuous washing for several days in running water to cleanse it from excess of acid, is saturated with a solution of saltpetre, which increases the expansion of the gases evolved by combustion.

By Clark's patent method, pyroxylinised wood grains, without been subjected to frequent washings, which always leaves some trace of acid, are combined with other constituents, which neutralise every particle of free acid. By this process the grains are immediately converted into an explosive of a character similar to black sporting powder.

All the elements which constitute pyro-pulver being chemically pure, combustion at the moment of ignition is instantaneous, leaving no residuum. Some care is required in loading the cartridges with this powder. The makers, I believe, prefer to supply them properly loaded by some mechanical contrivance, to ensure regularity in the shooting. This explosive, like gun-cotton and gun-felt, is not yet quite under control, being far too sudden in its combustion. If this, as we said before, could be remedied, it would be the great rival of black gunpowder.

As we gave a fair and impartial trial of gun-cotton, so we did the same with Schultze's Powder, and published the result in the *Field*, which we now give, so that the reader may judge of both for himself—

TRIAL OF SCHULTZE'S SAWDUST POWDER AGAINST BLACK GUNPOWDER.

SIR,—I am so often written to by sportsmen to give them my opinion on Schultze's Powder, that I decided to give it a careful and impartial trial against Lawrence's No. 3 grain, and, with your kind permission, to publish the result in the *Field* for the information of your readers. The respective numbers of each shot for pattern and penetration are given, with the average bracketed :—

	Schultze's Powder. Pattern.	Average	Schultze's Powder. Penetration.	Average	Lawrence's Powder. Pattern.	Average	Lawrence's Powder. Penetration.	Average
Four shots from central fire breech-loader, 12 bore. No. 1.	156	130	17	23	133	128	30	25
	125		22		121		23	
	131		24		130		17	
	149		28		129		30	
Ditto. No. 2 ...	137	127	20	22	121	124	21	24
	157		23		130		30	
	135		18		122		23	
	81		29		116		25	

I also tried it with one of our double rifles at 100 yards, with Lawrence's, four drams, and a spherical ball. I put six consecutive shots in a circle of six inches. Schultze, with a charge equal to the above, shot very wild, and did not put six shots in a circle of three feet, the bullets striking the target low. We tried the penetration at one-inch elm boards nailed one inch apart Schulze penetrated three, Lawrence penetrated four and a half boards. With the Boxer Enfield, Schultze penetrated three and a half boards; black powder, five boards; at fifty yards. With shot-guns the pattern is quite as good with Schultze's powder, or a little better, but the penetration is not so good. The cartridges were loaded by Mr. Clark for the purpose of trial.

St. Mary's Works, Birmingham. W. W. GREENER.

We have also given the pyro-pulver a careful trial, and now record the result. The average of six shots with a charge equal to 3 drams of black gunpowder and $1\frac{1}{8}$ oz. No. 6 shot was pattern 116, penetration 29. To make ourselves sure, we gave an extra trial loaded with a little more pressure, with the same gun and the same charge. The result was as follows:—120 pattern and 31 penetration. This is undoubtedly very good shooting, quite equal to black gunpowder; the objection is that the recoil is very sharp and severe. We also tried it with less shot, with a view to reduce the recoil; but this made no pattern, the average of six shots being only 57 instead of 120, although the penetration was slightly improved. As good shooting may be got with this pyro-pulver properly loaded as with black gunpowder, but the recoil is increased accordingly. It might be preferred by some for covert shooting, where the smoke of black powder is objectionable; but for any kind of rifle it is dangerous and quite unsuitable, being too quick in combustion. Several rifles, to our knowledge, have been burst from this cause with gun-cotton, and we consider this explosive partly of the same nature.

THE NEW PEBBLE GUNPOWDER.

The manufacture of the new "pebble" gunpowder, for our great guns, is now proceeding as fast as the resources of the Government factory will permit. Large supplies will also be obtained from the private trade. In appearance it is like black pebbles, whence its name. Its combustion is so much less sudden than that of the existing powder, that the maximum strain on the gun, with equal charges, is greatly reduced; and with larger charges the strain is far less than at present. This permits of the charges

Fig. 39. — One grain of pebble gunpowder.

of all our heavy guns being greatly increased, with a corresponding increase in the power of the weapons. The maximum charges will now, we believe, be fixed as follows:— For the 7-inch guns, 30 lbs.; 9-inch, 50 lbs.; 10-inch, 70 lbs.; for the 25-ton, 11-inch, and 12-inch guns, 85 lbs.; and for the 35-ton guns (11·6 inch calibre), 120 lbs.; these being the greatest charges which the guns are capable of consuming with advantage. As the strain due to these charges is, as we have stated, far less than with the smaller and less effective charges of rifle large-grain powder hitherto in use, it is clear that the benefits resulting from the introduction of the pebble powder will be very considerable. This large-grain powder was advocated by W. Greener in his work "On Gunnery," 1858.

THE NEW EXPLOSIVE DYNAMITE.

The Prussians used it during the campaign in blowing up bridges and buildings, but it seems better adapted for charging shells, and destroying an enemy's guns when you have obtained momentary possession of them in a sortie. Dynamite has 7½ times the strength of gunpowder, and is the safest to handle of all known explosives.

Heat alone will not explode dynamite, nor will concussion alone; if struck with a hammer on an anvil, the portion struck takes fire without inflaming the dynamite round it. A case containing 8 lbs. of dynamite (equal in effect to 60 lbs. of gunpowder) was placed on a brisk fire, and consumed without noise or shock; a similar case was thrown from a height of 60 feet on to a hard rock, without producing the slightest explosion; and a weight of 2 cwt. was let fall 20 feet on a case of dynamite, which it smashed, still without explosion. It can only be exploded by the combined action of the heat of a spark and percussion acting at the same moment of time, by means of a metallic capsule containing a strong fulminating composition, and fired by a slow match or the electric spark. For shells intended to pierce armour plates, or for extreme ranges, an explosive seven and a-half times as strong as gunpowder—and even safer in use—is a manifest *desideratum*. For such purposes the shell must be made so strong, that the cavity left for the charge has too little capacity to be very effective, unless it is charged with something a great deal stronger than gunpowder. Dynamite is unaffected by damp, and in burning or exploding it yields no smoke.

MR. GALE'S METHOD OF RENDERING GUNPOWDER NON-EXPLOSIVE.

It will be remembered that some years ago Mr. Gale invented, or revived, a process for rendering gunpowder non-explosive by mixing it with three times its volume of powdered glass, and that when so treated the powder when ignited burnt away quietly, without exploding. The result was curious, but to all practical men it was obvious that the invention was useless, for it involved fourfold storage room, and every charge had to be sifted (to free it from the powdered glass) before being used. But Nobel has applied an analogous process to nitro-glycerine with perfect success; he mixes nitro-glycerine with one-quarter of its weight of very fine sand; the result is a pasty powder resembling moist sugar in appearance and consistency, and quite as safe to handle. You may hold a cartridge of it in your hand and set fire to it with impunity; it merely burns away quietly, with a flame resembling that of spirits of wine.

THE PARACHUTE SHELL.

This is a kind of explosive shell, which will be used in modern warfare to illuminate. The following report will fully explain the principle:—

Some experimental trials have been carried out by the Royal Engineers on Chatham lines, for the purpose of testing the value of Parachute shells to light up a large extent of country.

Three of the large Parachute shells were sent up from Prince Henry's bastion, two of which exploded when at a great height, lighting up, by means of the magnesium light, the whole extent of Chatham lines, every portion of which was distinctly brought to view, and the various disposition of troops on the ground clearly discernible.

BREECH-LOADING BALL-GUNS.

Breech-loaders are much better adapted for ball shooting than muzzle-loaders. One great advantage of the breech-loader is that the bullet can be made to take the bore of the barrel tighter, on account of its entering at the breech instead of having to be rammed down from the muzzle; it consequently fills up the bore of the barrel more thoroughly, and better results are obtained. A ball gun, when properly made, will shoot up to fifty yards nearly equal to a rifle. Beyond that distance, a rifle is required for accurate shooting, as in a ball gun the bullet drops rapidly over fifty yards, and cannot be relied on. We give a diagram of the shooting of a breech-loading ball-gun at fifty yards, point-blank range—weight of the gun, 8½ lbs.; 12-bore, 28-inch barrels; charge of powder, 4 drams, No. 3 grain. Six shots consecutively fired at a 6-inch bull's-eye will put in five shots, and one, say, six inches to the right. We have tried many ball-guns, and give this diagram as a fair average of what may be done by a good-shooting one. We have also tried the same description of guns at 100 yards, aiming point-blank, and find the bullets drop about thirteen inches. The two stars at the bottom of the diagram indicate the fall of the bullet between 50 and 100 yards; the squares on the target represent six inches.

Ball-guns of this weight will do for shot, but are considered too heavy. Guns of 8 lbs. weight will do for ball, with $3\frac{1}{2}$ drams of powder, with about the same accuracy; $7\frac{1}{2}$ lbs. is a very suitable weight for a strong shot-gun, and will do also for occasional ball. This weight allows the barrels to be made stronger at the muzzles, which is essential

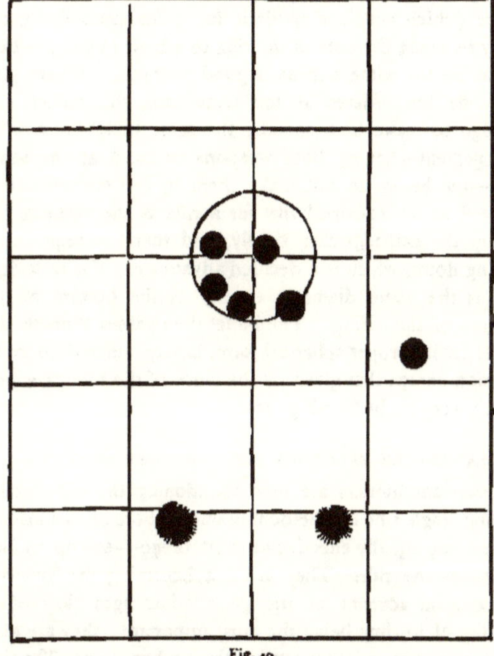

Fig. 40.

for ball-guns. The charge for this weight gun should be 3 to $3\frac{1}{2}$ drams of very coarse-grain powder. Its capabilities as a shot-gun are not impaired by using ball. The light ball-guns are not quite so accurate as the heavier, but at thirty or forty yards they are equally effective.

Sportsmen often desire their ball guns to be made to take the same size ball as that used in a No. 12 rifle, which is equal to a No. 11 gauge. There is a little difficulty here to be overcome, and it can only be accomplished in the following manner—by boring the barrels of the gun to No. 11 gauge (which certainly spoils it for a first-class shot-gun), or by reducing the bore of the rifle to a bare 12-gauge—that would be the same size as a good shot-gun. Where gun and rifle are ordered at the same time, the maker can manage to make both to take the same bullet by a little arrangement—having both weapons in hand at the same time—but he would not make them in this manner unless ordered, as an 11-bore bullet for a rifle is the best size; it fills up the cartridge-case tightly, and therefore requires no turning down, which is a decided advantage. The cartridge-case is the same diameter exactly as the bottom of the grooves of the rifling. The bullet then passes through the barrel in its proper spherical form, having suffered no compression except that given by the lands of the riflings, which is only very slight (see Fig. 42).

LARGE-BORE BALL-GUNS FOR ELEPHANT SHOOTING.

Elephant hunters are now abandoning the old muzzle-loading single rifle for the double smooth-bore breech-loader, these being equally effective at short ranges—say up to fifty yards—as the rifle. They are fast becoming the favourite weapon, on account of the great advantages they offer, rapidity of loading being the most important; they are also much more convenient when hunting on horseback. The size used is generally 8-gauge, with spherical ball, and 6 drams of powder; weight of gun, about 11½ lbs. More powder could be used by having the guns made a little heavier. 10-gauge guns are also used to carry 5 drams, and weighing 10 to 10½ lbs. A short, explosive, conical shell can be used with these breech-loaders to great advantage. A sportsman of our acquaintance assures us that he never has lost an elephant when using a shell from a 10-bore rifle. When struck in the head, the animals would instantly fall upon their knees, and then were easily despatched by a 16-bore. Some sportsmen advocate the small-bore Express rifle, with a solid hardened bullet, using a large charge of powder, to penetrate so as to reach the brain. But these bullets do not in all cases reach the brain, and then the force would not be sufficient to stop the animal; therefore we consider the sportsman has a far better chance of killing if he can stop him with the large bullet or shell, and prevent the creature from charging him. This plan gives the sportsman time to despatch him.

THE ORIGIN OF RIFLING.

Barrels were first grooved or rifled at Vienna, about the year 1498. The original object of grooving or rifling the barrels was to find space for the reception of the foul residue produced by discharging the rifle, and thus to diminish the friction of the bullet as it was forced down by the ramrod. During the next twenty years a spiral turn was given to the groove, and bullets were used with projections to fit the grooves, the degree of twist or spiral varying as the fancy of the gunmaker might suggest.

Many rifles are still grooved on the original plan, without spiral, for the Cape farmers, who use them with ball or shot.

SPHERICAL BALL RIFLES FOR LARGE GAME.

Indian sportsmen are now pretty well convinced that the spherical ball rifle is the most deadly weapon for large game shooting. There can be no doubt that, with a large charge of powder, rifles of 12-bore, and a spherical ball, will be found to stop a tiger or bear better than any conical bullet of the same size, as more powder can be used with it, and it is a much better bone-smasher. Besides making a more fearful wound, the shock given to the animal is much greater; the conical bullet wound seems to close up after the bullet has passed through the fleshy part, not letting out the blood so freely as if caused by the spherical, and causing much less shock to the animal. This is the reason why we advocate the large spherical ball. The above description refers to the solid conical bullet, not a conical shell or hollow-pointed bullet, as these expand when striking, and make a frightful wound.

Lieut. Forsyth, in his work, "The Sporting Rifle and its Projectile," lays down very clearly the form of rifling and spiral that is best adapted for this kind of rifle.

The spiral varies according to the size of bore; in a 16-bore one turn in 6 or 7 feet would be necessary to give the bullet an accurate flight. The larger the bore, the less the spiral required. 8 ft. 6 in. turn is suitable for a 12-gauge; 10 ft. for a 10-gauge; 12 ft. for an 8-gauge. If more twist is given to the rifling than I have above mentioned, it is superfluous, and loss of power. Rifles of this class are made with a much quicker turn of rifling, but they are intended to throw a spherical and a conical bullet out of the same barrel. We consider it a mistake to have a quicker spiral than 8 ft. 6 in. for a 12-bore, as a short, blunt, conical bullet can be effectively used with this rifling. We give an illustration of this bullet with the rifle, Fig. 41.

This short bullet, with 3 drams of powder, will strike the same object at 100 yards as a spherical ball with 4 drams, and is quite as accurate with the same sighting. There are projecting rings on this short conical bullet that fill up the grooves of the rifling, making it mechanically fitting. It does not expand. The breech-loading system admits of a very tight bullet being used, which, with a thick felt wad, prevents effectually the windage, and no force is

Fig. 41.—Spherical Ball Rifle. No. 10 gauge.

lost. The tight-fitting bullet gives a flat trajectory and a long point-blank range.

THE BEST KIND OF GROOVING FOR SPHERICAL BALL RIFLING.

Opinions differ among first-class gun-makers as to which is the best form of grooving, and each has his own favourite plan. We give illustrations of three kinds that are ex-

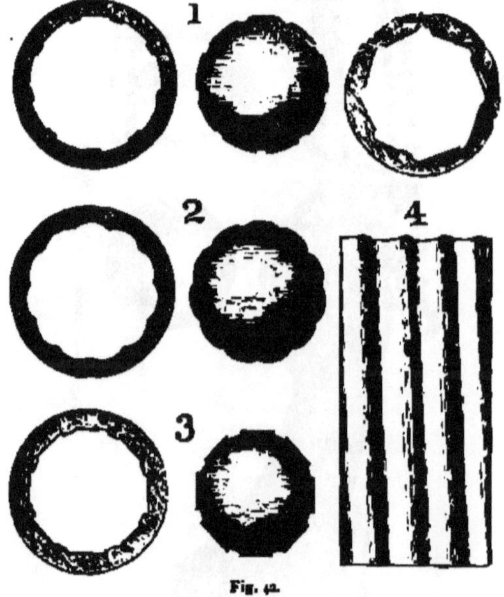

Fig. 42.

tensively used. Sketch No. 1 is the form we prefer. The bullet retains its original shape better after having passed through this shallow rifling. The twist or spiral being very slow, and the bullet fitting very tight, very shallow grooving will suffice. These bullets never strip. The number of grooves does not matter; nine or eleven will answer well. This we call "shallow rifling."

No. 2 illustration shows a form of rifling with the grooves rounded, leaving no sharp angles. The object of this is that it is easier cleaned. The grooves appear shallow, but in reality are deeper than No. 1. Our objection to this kind is that it cuts away the outer edge of the bullet and destroys the sphere.

No. 3 illustration shows the "Henry" form of rifling. It is open to the same objections as the No. 2 we have above described for spherical bullets, but for conical expanding bullets it is perfection.

Rifling should be arranged to suit the particular kind or form of projectile that is intended to be used, and the dimensions of the projectile should be determined according to the range desired.

The smooth-bore, we say, has a point-blank range of 50 yards with spherical ball—by rifling the same barrel we increase the range to about 85 yards, with the same charge of powder—and is tolerably flat up to 180 or 200 yards, but after that distance it drops suddenly. To get a longer range, the bore must be reduced, retaining the same weight of projectile; consequently, it must be elongated, as this form offers less resistance to the atmosphere. This long bullet, to be kept point foremost, must receive a good spiral. For a 25-bore, about four feet is necessary. That would travel 1,400 yards, even with a less charge of powder than is used with a 12-bore spherical. By reducing the bore to 50, and increasing the spiral to 1 in 22 inches, a range of 2,000 yards can be obtained.

A small-bore conical bullet with a very sharp spiral is much slower in its flight than a large spherical ball and a slow twist; but the trajectory of the spherical bullet up to

200 yards is considerably flatter, and, therefore, much better adapted for large game. Should the sportsman miscalculate his distance some twenty or thirty yards, this rifle would give him a much better chance of hitting than a long-range rifle.

We are prepared to state that a properly constructed rifle will give a strictly point-blank range of 85 yards. At 100 yards the fall would be very slight, so slight, indeed, that it would not affect the aim even up to 120 yards. The bullet drops very little up to 150 yards. Many sportsmen use no back-sights up to this distance, shooting from the rib, as with a shot-gun. A rifle made as above described would not miss an object at 15 or 20 yards. An artificial point-blank may be given to a rifle by raising the rib, at the breech end. The objection to this would be that the bullet would rise at a short range, and probably miss the object, by shooting over the mark, but would appear to make a longer point-blank.

It is essential that the body of the breech-action should be made very long, and the hinge-pin a good distance from the standing breech. If made in this way, and with back-action locks, there is no drooping of the barrels at the muzzle. We consider the double-grip principle of action is the strongest, and gives the greatest binding power.

We have heard of double rifles having only 40 yards of point-blank range. This is caused by a weak action, which allows the barrels to droop at the muzzle at the moment of firing.

Breech-loading rifles built on the plan we have laid down will have a low trajectory, and shoot equal to a muzzle-loader, which we have proved by numerous experiments.

THE ACCURACY OF DOUBLE RIFLES.

Lieutenant Forsyth, in his work called "The Sporting Rifle and its Projectile," gives the extreme sporting distance as 150 yards for large game shooting, and most game is bagged at from 80 to 100 yards.

A really good shooting spherical ball rifle is as accurate a weapon as can be constructed, taking into consideration the large-size bore, viz., 12 or 16-gauge. No conical ball rifle of the same bore can shoot better, or will admit of so large a charge of powder being used.

The conical ball is acted upon more than the spherical by a strong wind crossing the line of flight. It is remarkable how little the spherical ball is acted upon by the wind. We have made splendid practice with spherical ball rifles on very windy days. We always use a table-rest when trying our rifles for accuracy, with a sand-bag to rest the rifle upon.

We give an illustration of a target divided into squares of 6 inches, with a 6-inch bull's-eye for 100 yards.

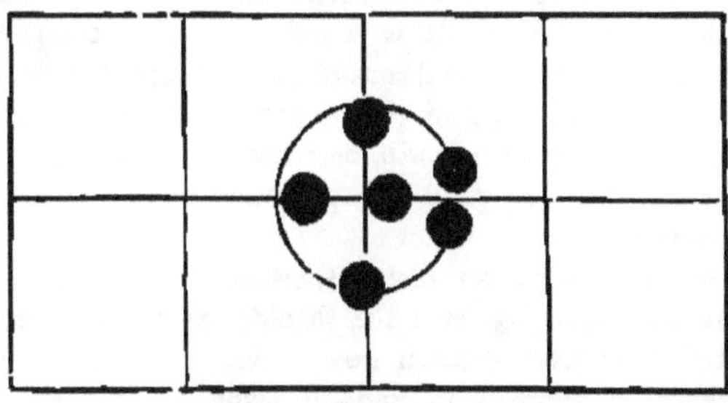

Fig. 43.

We put in a 6-inch bull's-eye six shots consecutively, with a first-class rifle, at 100 yards; this would be a mean deviation of less than 2½ inches from the centre of impact.

With these rifles at 150 yards the shooting is very good; beyond that distance they begin to fall off rapidly, and more particularly in light rifles. We find the best charge of powder for these rifles of 12-bore is 4 drams of Lawrence's No. 3 grain, and 3½ drams for the 16-bore. More powder can be used for game shooting at close quarters; but if more is used it will affect the accuracy slightly. We mention this because, when match shooting at a target, it is an important point.

Many sportsmen are in favour of the very coarse grain powder for rifles. We believe in it for conical bullets with sharp spirals. We find No. 3 is better and gives greater accuracy than No. 4 for spherical balls. The bullet must have a very high rate of velocity given to it, or it loses its accuracy and great striking power. No. 3 grain powder is quicker, and consequently better adapted for this bullet.

We make single-barrel rifles of 12 and 16-bore for large game shooting, but cannot get any better shooting from them than we now get from the double-barrels. This shows the improvements that have been made in the construction of the double rifles. It is stated by a very competent person that a really well-constructed and accurate double rifle is worth its weight in gold to the Indian sportsman; I mean one that will shoot with both barrels like a single, and make a diagram as good as represented by the preceding engraving.

We do not expect that a sportsman will be able to make such shooting from the shoulder as we have represented. We have done it from a rest repeatedly, in the presence of sportsmen; and all first-class rifles will do the same, from a rest, when directed by a really good shot.

We gave as a prize to the National Rifle Association of India one of our best-shooting rifles, one that would shoot as above represented, viz., put six shots in a 6-inch bull's-eye at 100 yards. Although won by one of the finest shots in India, the best diagram made from the shoulder at 100 yards was about equal to our shooting from a rest at

200 yards' range. We introduce the account, taken from the National Rifle Association of India's reports, as we think it may perhaps be interesting to some sportsmen. We never allow rifles to leave our hands that will not put six shots consecutively in an 8-inch bull's-eye at 100 yards.

The following account of a match for a Greener Presentation Rifle, shot for at Nynee Tal, which appeared in the local papers, is published as a postscript to the National Rifle Association of India's report; it may prove interesting to many sportsmen:—

"Mr. W. W. Greener, gun-maker, of St. Mary's Works, 61 and 62, Loveday Street, Birmingham, sent out to Mr. Currie, to Nynee Tal, to be competed for by not less than ten competitors, one of his double-barrel breech-loading central-fire 12-bore rifles, weight 11 lbs., price £35 (case and fittings, £3 10s. extra), on the following conditions, viz., that the match should be shot for at 100 yards, every competitor using the Presentation Rifle, and also two other similar weapons, by other known makers, and that the competitor who made the highest aggregate score with all three weapons should take the Presentation Rifle.

"The match could not be shot off in one or two days, and extended over a week, from 29th September to 6th October, as there were no less than sixteen competitors—2nd and 3rd prizes being arranged for out of the entrance money.

"The Hon. C. Dutton lent a 12-bore breech-loading central-fire Rigby, and Mr. Currie lent a 12-bore breech-loading central-fire Turner; and, as will be seen by the subjoined score, the shooting of all three weapons was very even. A spherical ball and 4 drams of powder (and, in some instances, 4½ drams for the Rigby) was the charge for all the weapons.

"The Presentation Greener made the best shooting, for, in the aggregate made by all ten competitors who fired through the whole match, the Greener's score is 220 against the Turner's 214, and the Rigby's 210; and if the score of only the three first, viz., the three prize-winners, be taken, the Greener still retains the first place, the scores being Greener 74, Rigby 72, Turner 69. Two sighting shots with each weapon were allowed to every competitor, and they all competed on as even terms as it was possible to make. It is but common justice to Mr. Currie and his rifle (Turner's) to state that he had never used the charge of 4 drams of powder and a spherical bullet with his weapon, and did not know how it shot with that charge; the regular charge for the weapon, and that always used by him, being 3 drams of powder and a solid conical or else a hollow shell projectile. Mr. Dutton's is a new rifle, and he has had very little practice with it at the 100 yards or any other range, so that the owners of the two rifles cannot be said to have had any unfair advantage; and, strangely enough, they both made better scores with each other's weapons than with their own.

"Mr. W. W. Greener guarantees to send out rifles of 12 or 16-bore for £35 (case and fittings extra from £3 10s.) to order, as good in every way as this weapon, which has beaten two other good weapons in this match, and which undoubtedly is a very accurate handy rifle, on his special patent self-acting-strikers principle, which may be seen advertised and illustrated in the *Field*.

"The competition was particularly good and close, and Mr. Ross is the winner, with 73 points; Colonel Cuppage and Mr. Currie tying for second place, with 71.

"The ties were shot off on October 7th, with the Presentation Rifle, from the shoulder, at 100 yards, according to the conditions previously laid down for decision of ties. Colonel Cuppage beat Mr. Currie, thereby gaining the 2nd prize.

"The target used was the National Rifle Association of

India's special double-rifle target, viz., the bull's-eye, six inches square, scoring 4; the inner centre, one foot square, scoring 3; the outer centre, two feet square, scoring 2; rest of target, six feet by four feet, outer, scoring 1. Eight shots with each weapon, viz., four standing and four in any position, at 100 yards.

NAMES OF COMPETITORS.	RIGBY 12-BORE C. F. WEIGHT 10½ lbs.		TURNER 12-BORE C. F. WEIGHT 10 lbs.		W. W. GREENER 12-BORE C. F. WEIGHT 11 lbs.		GRAND TOTAL.
	Stdg.	Any Pos.	Stdg.	Any Pos.	Stdg.	Any Pos.	
1. H. G. Ross, Esq., C.S.	12	14	12	11	11	13	73
2. Lieut.-Col. Cuppage, C.S.	11	10	11	12	11	16	71
3. R. G. Currie, Esq., C.S.	12	13	11	12	9	14	71
4. A. M. Markham, Esq., C.S.	9	13	7	14	12	14	69
5. Major Anderson, C.S.	11	13	10	8	13	13	68
6. S. Berkeley, Esq.	11	9	13	10	10	9	62
7. General Story, C.B.	8	10	11	8	12	12	61
8. Hon. C. Dutton	12	9	10	12	7	10	60
9. C. C. Macrae	7	12	9	12	9	7	56
10. T. J. Ryves, Esq.	8	6	10	11	8	12	55
Totals	210		214		220		

"There were six other competitors, who withdrew at different stages of the competition, and did not complete the match. "ROBERT G. CURRIE, C.S.

"*Nynee Tal, 9th October*, 1869."

THE LOADING OF CARTRIDGES FOR DOUBLE RIFLES.

The chamber in a double rifle is usually made to take the cartridge-case full length, which is 2½ inches. It is most important that the cartridge-case should not be longer than the chamber, even one-sixteenth of an inch. When it is so, the cartridge-case gets contracted just at the edge, and is turned inward; the result of this is that the bullet is stripped and reduced quite one size passing from the chamber into the barrel, and then cannot properly take the grooves of the rifling, causing the rifle to shoot wild. A sportsman would get cartridge-cases the exact size, if ordered from the maker who built his rifle, as it is usual for gun-makers to gauge all the cartridge-cases sent abroad with their rifles. Metal-lined cartridge-cases are best for large charges of powder; the green "gas-tight" are the next best; but blue or brown should not be used. The usual charge of powder, with a thick felt wad, and then the bullet, will not quite fill up the cartridge-case; this is of no consequence, and will not affect the shooting. The cartridge-case must not be cut shorter than the chamber of the rifle. Should the bullet be loose for the case, coat it with bees-wax, and turn down the cartridge-case. No wad must be placed over the bullet, or a burst barrel may be the consequence.

The bullet should fit the cartridge-case tightly, and the case is better not turned down. The wad should be of the very best hard thick felt. A paper wad over the powder is useful, to prevent the grease from the lubricated felt wad reaching the powder; and a re-capped cartridge-case should not be trusted when in pursuit of dangerous game.

THE WEIGHT OF DOUBLE RIFLES.

The usual weight of a 12-bore double rifle is 11 lbs. This allows the barrels and actions to be made strong enough to insure accuracy and durability. Of course, they can be made as light as 8½ lbs. with perfect safety, to suit the requirements of the sportsman; but we would not recommend a rifle lighter than 10 lbs., if the sportsman is at all equal to that weight. If a very light rifle is necessary, then we should say have a 16-bore, 8 lbs. weight, that is 1 lb. lighter than is usually made for India. A double rifle of 9 lbs. weight, 16-bore, will shoot equal to a 12-bore of 11 lbs. weight.

The length of barrels preferred is 25 inches; which is sufficiently long for all practical purposes. Light rifles may, however, be reduced to 22 inches with advantage. The barrels should be either of laminated steel or plain steel. The former is preferable for very light rifles.

Fig. 44.

DOUBLE RIFLE, 577-BORE, FOR SNIDER CARTRIDGE, ON THE LEFAUCHEUX PRINCIPLE.

The engraving represents a useful double rifle, taking the regulation cartridge; the weight is only 8½ lbs. It is rifled with five grooves, with one turn in 4 feet, the same as the short Enfield. It shoots well up to 300 yards; and throws a spherical ball of twenty-

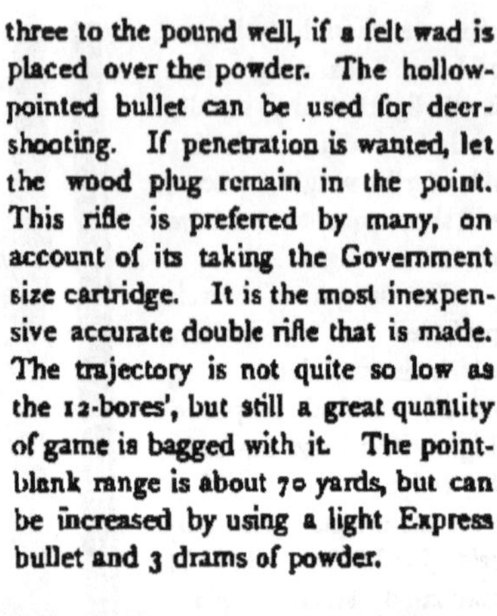

three to the pound well, if a felt wad is placed over the powder. The hollow-pointed bullet can be used for deer-shooting. If penetration is wanted, let the wood plug remain in the point. This rifle is preferred by many, on account of its taking the Government size cartridge. It is the most inexpensive accurate double rifle that is made. The trajectory is not quite so low as the 12-bores', but still a great quantity of game is bagged with it. The point-blank range is about 70 yards, but can be increased by using a light Express bullet and 3 drams of powder.

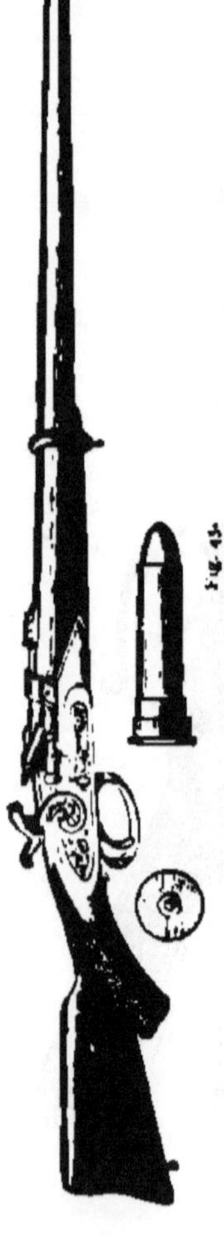

Fig. 45.

SINGLE SPORTING SNIDER, 577-BORE.

This is an excellent little weapon; and is very handy. The barrel is 26 or 28 inches long. It will shoot well up to 300 yards; a good one will shoot equal to an Express for accuracy, but it has not so flat a trajectory. The engraving shows that this rifle is, like a musket, fastened to the stock by a band; it is made also without a band. The barrel can be taken easily from the stock. To load, the hammer must be raised to half-cock, the block must be brought over by the thumb from left to right; the extractor is attached to the breech-block. By pulling the block backwards the cartridge-case is extracted. There is a spiral spring that takes the block back into position. The latest improvement in this rifle is the spring-bolt, which keeps the breech-block in position, and prevents it from being forced open by a defective cartridge. This weapon makes an admirable carbine for use on horseback, and is the only reliable rifle that can be supplied at a low price. The usual weight is 6¼ lbs., but it is best if made 7¼ lbs.

SPORTING RIFLES, 500-BORE.

The bore of these rifles is about 36. They are much more accurate than the 577 at long distances. Being larger than the 450 Express, they are often preferred to the Express rifle. They are made with a quick spiral, about one turn in 30 inches. The bullet is hollow at the base for a plug; it is also hollow at the point, to be used with or without a plug, the same as the Snider. This rifle takes a charge of powder 85 grains. The trajectory is rather high, but it is very accurate, especially at from 300 to 500 yards. Although only intended for deer-shooting, it is often used, and with great success, for tigers, bears, and such like game; but in such cases it is best to have a large bore in reserve.

This 500-bore is made also to take 4½ drams of powder, by increasing the length of the cartridge-case. The bullet is shortened, and the rifling is slower in the spiral; the effect of this is to increase the point-blank range, which it does very considerably. It strikes with a greater force, and the bullet expands very much, making a formidable wound. The point-blank range with 4½ drams of powder is about 130 yards. It cannot be relied on for extreme accuracy over 200 yards; but at 140 to 160 it is very accurate.

THE EXPRESS RIFLE, 450-BORE.

This is a small-bore rifle, about 52-gauge, taking a large charge of powder, and having a very flat trajectory. It has been causing some stir among sportsmen for deer-shooting at long ranges. The actual point-blank range of this kind of rifle, with four drams of powder, is 130 to 140 yards. The bullet is a short cone, with a hollow point. The hollow in the bullet causes it to expand on striking, thereby inflicting a most fearful wound. The bullet gets flattened to about ten times its original size. Another advantage obtained by the hollow is that the bullet can be made lighter, and yet retain the desired length, to take the rifling properly. The larger charge of powder with this light bullet gives a very high velocity, and long point-blank range. Many letters have recently appeared in the *Field* newspaper, from Indian

Fig. 46.

sportsmen who have used these rifles at long distances at antelope and small deer, with great success. One sight is only necessary for sporting distances; and over 200 yards a full fore-sight must be taken, as the bullet begins to drop at over 150 yards. This is undoubtedly the best kind of rifle for hill-shooting, where it is difficult to judge distances. Animals shot at ranges varying from 50 to 140 yards must be struck if the aim is correctly taken. But for large and dangerous game shooting a much more powerful rifle is necessary. For such purposes we should recommend spherical ball or explosive shells.

The 450 Express rifle cannot be relied on for extreme

accuracy over 200 yards. Long range is not obtained with this kind of rifle and light hollow bullet. Distance of flight is sacrificed for the desiderata—flat trajectory and long point-blank. These rifles are admirable for the purpose for which they are built, but would be utterly useless for South Africa, where sportsmen shoot at game at 500 yards or upwards, and expect to kill; but by using a solid bullet and less powder, the range may be increased considerably.

The weight of Express rifles, either double or single barrels, should be 9 lbs. to get good shooting. The weight is only required to counteract recoil. The rifle would be perfectly safe if made 7 lbs. weight, though it would slightly interfere with the accuracy, and very fine shooting could not be made with it. We get splendid shooting from rifles of this class weighing only 8 lbs., when shooting from a rest.

The Martini-Greener sporting rifle, single barrel, makes a splendid Express. It takes a long brass coil cartridge-case,

Fig. 47.—The Sporting Martini-Greener Rifle.

containing a charge of four drams of powder. We introduce an illustration representing rifle and cartridge-case at page 117. The breech action is exactly the same as that approved of by Government, but the barrel is shortened to twenty-eight inches: the spiral is considerably reduced. The manipulation of this rifle is very simple: By depressing the lever sharply the empty case is jerked out over the shoulder. This motion cocks the piece. By placing in a loaded cartridge and pulling up the lever again, it is ready for firing. There is a safety bolt, which secures the trigger; also an indicator, to show when the piece is cocked. We shall describe the mechanism of this weapon more fully when treating of military rifles.

THE WESTLEY-RICHARDS SINGLE EXPRESS RIFLE.

This rifle is similar in construction to the Martini. It is considered a great improvement on that system. The breech block is exactly the same, working on a pivot at the rear. The block is depressed to admit the cartridge by the lever over the trigger-guard, which cocks the rifle and extracts the cartridge-case at the same time. The arrangements are very simple, there being only twelve parts and seven pins. It has a powerful cartridge extractor, forcing the case clear of the rifle. The main-spring is not a spiral, but one of the ordinary kind. The nose of the hammer, similar to that of a revolver, strikes the cap itself, there being no needle or exploding pin required. We give a representation of this rifle at page 210, Fig. 107.

The action is just the same as that of the military rifle, but is made to take a larger cartridge-case, 2¾ inches long, containing 120 grains of powder, and an Express bullet of 260 grains. We introduce a sketch of this cartridge, full size, at page 124, Fig. 50.

This cartridge is certainly the best that has yet been introduced for the Express rifle, being far superior to the brass coil. It has a brass-drawn case, with a solid base. It stands the large charge of powder admirably, and can be used over and over again. It expands after a few shots have been fired; but a swedging tool is sent out with the rifle, to compress the case to its original size, and by this plan being adopted the case can be used twelve or fourteen times. This must be a great advantage to Indian sportsmen in localities where cartridge-cases are difficult to be procured; but the great feature in this rifle is that the large charge consisting of 120 grains of powder can be used, which gives increased point-blank range—viz., 150 yards—with tremendous striking power. The weight of the rifle, single, is 8½ lbs. to 9 lbs. The recoil is not excessive; this is owing to the light bullet. We consider this rifle one of the best of the kind that is made. One great point that will be appreciated in this weapon is its safety. It can be loaded or unloaded with the trigger bolted. The shooting is good. (See Wimbledon report.)

The Westley-Richards rifle takes a lighter bullet than other Express rifles. The usual weight is 320 grains for a 450-bore. The advantage gained by the light bullet is the extremely long point-blank; but over 200 yards the shooting is wild. A bullet weighing 320 grains will be tolerably accurate up to 230 yards. The plan adopted by some makers of putting only one sight for all distances, we do not approve of. There is a perceptible fall of the bullet from 150 to 200 yards. We prefer one standing sight for the point-blank, and one leaf to give the 200 yards. We think this is preferable to making allowance by taking a full fore sight.

The Soper breech-loading arrangement also makes a good sporting Express rifle. The mechanism is simple, strong, and durable, and can be loaded rapidly. The lever is placed at the side of the lock, in a very convenient position. The extractor is also very powerful. This rifle can be made to take the brass-drawn cartridge-case—4½ drams of powder. We give a drawing of this rifle in the military section.

All these small-bore rifles are usually rifled according to Henry's principle. Very accurate shooting can be got with this description of rifling. There are other forms equally as accurate for sporting distances.

The Henry breech-loader is much admired by many sportsmen. It retains the ordinary lock and cock. The principle of the breech action is the same as that of the well-known Sharp's American breech-loader, but it takes the Boxer cartridge. The lever fits over the trigger-guard, instead of the lever and guard being in one. Sharp has recently altered his rifle to take the Boxer cartridge, or a similar one of brass-drawn. The action of Sharp's breech-loader is figured in the military section. It will be seen that the breech block slides down below the line of fire: the objection of this arrangement is that the exploded cartridge-case sometimes expands and fixes the block, which makes it difficult to open. The Martini or Westley-Richards plan is a decided improvement upon Sharp's. The block cannot possibly get jammed, as it works on a pivot at the rear. The front part of the block describes part of a circle, moving clear of the disc of the cartridge the moment it is depressed by the lever.

EXPRESS RIFLE FOR LARGE GAME SHOOTING.

Sportsmen seem pretty well convinced, from experience, that the 450-bore Express rifle is much too small in the bore to be thoroughly effective and deadly when used against large game, such as tigers and bears. The 500-bore or half-inch, which is correctly (by the gun gauge) 38-bore, is acknowledged to be the best size of Express rifle, for the above purpose. The cartridge for the 500-bore is shorter than that for the 450-bore, but contains 4½ drams of powder, being half a dram more: but this half-dram is not sufficient to compensate for the increased diameter of the bullet—hence the trajectory is not so low. 120 to 130 yards is the extreme point-blank range of the 500-bore. The striking power is, however, much

Fig. 48.—Single Express Rifle, 500-bore.

greater, in consequence of the increased weight of the projectile. It makes a more formidable wound; the bullet expands to a much greater size than the 450. The shock to the system is considerably more than would be imagined from

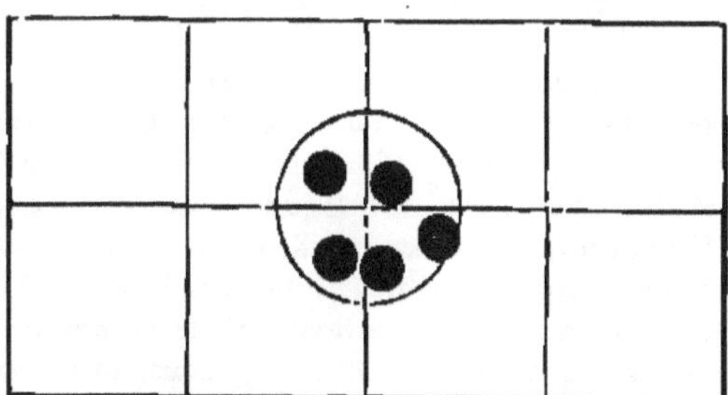

Fig. 49.—Target of Express Rifles at 100 yards.

the lightness of the bullet. This form of projectile is undoubtedly the correct one, as the whole force of the large charge of powder is expended upon the animal. The bullet is so flattened upon striking that it remains in the body.

The result is different with a solid bullet and a large charge of powder; the projectile generally passes entirely through the body, a mere waste of power, as it fails to give the same shock as the hollow-pointed bullet.

Solid conical bullets are used successfully in Express rifles upon small deer; in fact, this form is preferable to the hollow point in some cases, as that form of bullet mutilates and damages the skin. We have heard of this projectile carrying away part of the off side of a small deer. This solid bullet is used by some sportsmen even for elephants, where great penetration is required. We know of many instances where tigers and bears have been killed with one shot from a 500-bore Express rifle, and we feel confident that this weapon will become the favourite with Indian sportsmen. The spiral of the rifling is not so slow as some would imagine; on the contrary, it is moderately quick. This is necessary to keep the projectile point foremost; the base of the bullet being the heaviest, it has an inclination to turn over.

LARGE-BORE EXPRESS RIFLES IMPRACTICABLE.

Many Indian sportsmen have an idea that a large-bore Express rifle can be built with a trajectory as low as a small-bore. This may be accomplished, but the weight of the rifle would have to be so increased that no sportsman could carry it.

The principle of the Express rifle consists of a small bore and a large charge of powder. If the bore is increased, the charge of powder must be increased in the same ratio. A 500-bore would require a charge of 5½ drams to be equal to the 450 for trajectory. No cartridge-case is yet made to hold this charge. The rifle should be 12 lbs. weight to counteract recoil. For example, if the bores of these rifles are increased to 577, the Government size, they would require 7½ drams of powder to give the same trajectory as the 450, and the weight of these rifles should be 15 lbs. It must be borne in mind that there is not only the increased weight of lead to be projected, but a greater column of air to be displaced, from the increased diameter of the bullet. Lieutenant Forsyth endeavoured to overcome this difficulty by making a tubular projectile. So far as trajectory was concerned he was successful, but the projectile was found to be very deficient in accuracy of flight. These bullets could be made very accurate by machinery; the hollow must be made perfectly true, and an equal thickness of lead on all sides of the tube. Should there be the slightest inequality, it would cause a great divergence of the projectile, on account of the spiral given by the rifling and the air rushing through the centre of the bullet. This would give an excentric motion to it, which would spoil all accuracy. It is almost an impossibility to cast a perfect bullet of this kind from an ordinary mould. The tubular bullets are not likely to become very general for the reasons above stated.

CARTRIDGES FOR EXPRESS RIFLES.

Gun-makers have had considerable difficulty in procuring the kind of cartridge-case suitable for the Express rifle. Great credit is due to the Messrs. Eley Brothers, the well-known ammunition makers, for the readiness they ever evince to meet the requirements of the gun trade, by manufacturing special cartridges for the Express rifle. They have not, however, yet arrived at perfection. Many improvements have lately been made in the Boxer case, and there is still room for more. The 450 case, which is 3¼ inches long, holds barely 4 drams of powder. With an Express bullet this case answers pretty well, but is considered inconveniently long. The 500, or half-inch, is better: it is but three inches long, and contains 4½ drams. This is quite as much as the case will stand without splitting. We consider the brass coil case without paper covering the best.

There are cartridge-cases made on the Boxer principle "bottle shape" (Fig. 51), which will probably be adopted by the Government, and will answer for a military weapon, where the cartridges are supplied loaded.

We are not yet convinced that these "bottle-shape" cases are suitable for sporting rifles: they are inconvenient to load. The best "bottle-shape" cartridge-case is the

Fig. 50.—The Brass-drawn Cartridge.

Fig. 51.—The Brass-crimped Boxer Cartridge.

Fig. 52.—The Tin Boxer Cartridge.

Berdan "American." It is a brass-drawn case, made from a solid piece of metal. It is used very generally in American rifles, such as the "Remington" and "Peabody." Cartridges of this kind, however, contain too small a charge of powder for Express rifles. There are at present being made in Birmingham cartridge-cases of the above pattern, which will hold 4½ drams of powder for 450-bore: but the supply is limited, and likely to continue so for some time to come.

The machinery for constructing them is special, and the makers are full of orders for military cartridges. We should hesitate before sending out rifles, knowing the difficulty there would be in procuring suitable cartridges. We give a drawing of this case (Fig. 50): it is 2¾ inches long. It is very strong, and can be used many times by recapping, and slightly compressing it by a swedging apparatus, which is necessary, as the case expands slightly after each discharge. There is also a similar case made for 500-bore, which contains 5 drams of powder. We must content ourselves with the Boxer case for the present, until the brass-drawn can be obtained in larger quantities. There is, however, a "bottle" cartridge on the Boxer principle made of tin (Fig. 52), just introduced by Messrs. Eley, which will contain a charge of nearly 5 drams of powder for 450-bore. We believe the case will answer well for Express rifles.

REMARKS ON THE ACCURACY AND TRAJECTORY OF EXPRESS RIFLES.

We have been assured by a celebrated Indian sportsman that he has put six shots consecutively in a six-inch bull's-eye, at distances from 25 to 125 yards, using one sight, from a 500-bore double rifle. This, it must be admitted, is very accurate shooting; and agrees with that which we see done from the rest with these rifles.

We have another account from a sportsman who killed seven deer consecutively with a 450-bore rifle, at distances varying from 70 to 130 yards, two only taking a second shot. 130 to 140 yards is about the correct point-blank, with 4 drams of powder, of the 450-bore rifle: the bullet would drop only 3 to 4 inches at 170 yards, and still give a chance of game being struck at this distance with the point-blank sight.

Some gun-makers advertise their rifles as having a point-blank for 200 yards; but this is accomplished by raising the rib at the breech. These rifles will shoot high at short ranges. Sportsmen can prove this by shooting their rifles at a small object at 10 yards distance: we allude to rifles weighing only 9½ lbs., with a charge of 4 drams of powder. Rifles can be made to take the new tin bottle-neck cartridge, which holds 5 drams of powder; this will give a longer point-blank, but necessitates a rifle of 10½ lbs. weight to counteract the recoil; such a weapon would be considered too heavy for deer-shooting by many sportsmen. A rifle 500-bore weighing 11 lbs., with this cartridge and 5 drams of powder, would be a formidable weapon for tigers and bears.

THE ADVANTAGES OF THE EXPRESS RIFLE.

Long point-blank range is the principal advantage possessed by the Express rifle. The introduction of the small-bore for game-shooting is claimed by the American sportsmen. The old Kentucky rifle approaches as nearly as possible to the Express principle, which is simply the light bullet and comparatively large charge of powder. The modern Express is merely the substitution of the hollow-pointed bullet, which gives a larger striking surface, and consequently makes a much larger wound, caused by the expansion or flattening out of the bullet.

Sportsmen are well aware of the great difficulty of judging distances, especially when shooting over rough uneven ground. We introduce an illustration to assist our description. We suppose a sportsman shooting at a deer with a spherical ball rifle at 150 yards, with the 100 yards sight up. It is most probable that he would miss altogether, as the ball would drop rapidly over the 100 yards. If, on the contrary, he fired under the same circumstances and at the same distance with the Express rifle, he would most certainly hit with the point-blank or standing sight. On the other hand, if the distance could be accurately guessed, the spherical ball rifle would be equally effective. We give the drawing simply to show the trajectory of each rifle—the dotted line is that of the Express, which is good for 150 to 160 yards; the spherical ball rifle is correct enough to strike at 110 yards point-blank.

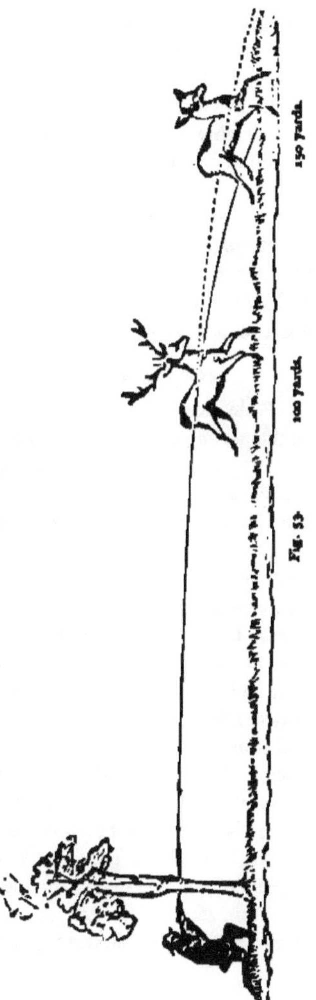

Fig. 53.

The double Express rifle, 500-bore, taking a charge of 4½ drams of powder, is the most useful rifle for every description of game-shooting where only one rifle is employed. It is accurate enough for deer-shooting at long ranges, and powerful enough for the destruction of more formidable animals. This rifle can be made 10 lbs. in weight or even less, and still be accurate and serviceable. An extra 1 lb.

will give greater steadiness to the weapon, if the sportsman can carry this weight. There is no advantage in having a single Express rifle except in cost. The Lefaucheux single rifle is most in favour; but the Martini is considerably cheaper, and shoots quite as well. The only objection to this system is the arrangement of the lock, which is considered by some persons to be dangerous. We are just now bringing out a single rifle on the Martini system of breech block, associated with an improved cock with the hammer outside, so placed that it can be cocked with the thumb. This weapon is specially intended for sporting purposes.

STEEL BARRELS FOR SMALL-BORE RIFLES.

Plain steel barrels are now used for all small-bore rifles, both sporting and military. The American mode of manufacturing them is to prepare a bar of steel the length of the barrel, and drill the hole from end to end. The English plan is to prepare a short thick bar of steel, say ten inches long, and drill or punch the hole through it; then heat it red-hot and pass it through grooved rolls, inserting a mandril or steel rod into the bore to prevent it from closing up; the rolls are made to draw out the barrels' taper, and to the required length. The barrels are then straightened, and undergo the process of boring and turning in the usual way. Barrels made in this manner are far superior to the old iron Enfield rifle barrels, as there is no weld or join in them. The quality of steel is mild, fibrous, and equal in temper throughout. The advantage of this description of barrel is, that it is not liable to honeycomb by the corrosive action of the gunpowder, which is a serious fault in barrels of a mixed metal, such as is used in sporting guns.

Plain steel barrels can be rifled more perfectly, and will retain their good shooting qualities longer, than those of a

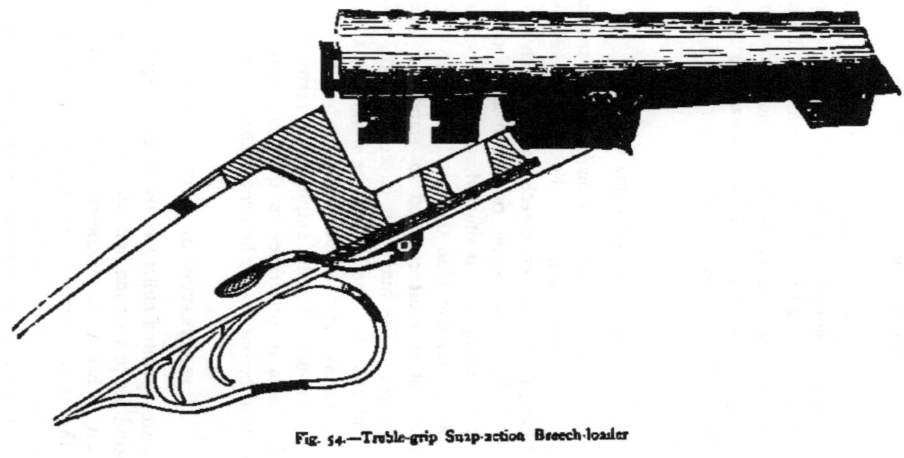

Fig. 54.—Treble-grip Snap-action Breech-loader

J

softer metal. They will likewise resist a very large charge of powder, and are perfectly safe to use for sporting purposes, either for large or small bores. They are also well adapted for heavy ball-guns, but we do not recommend them for light shot-guns. In appearance these barrels do not equal laminated steel or damascus, they show no figure, but resemble musket barrels.

DOUBLE RIFLE WITH SNAP-ACTIONS.

Many sportsmen prefer rifles made on the snap-action principle. We give at page 129, Fig. 54, a representation of a treble-grip snap-action breech-loader, which we have specially designed for double rifles and large-bore duck-guns. The locking bolt is situated immediately under the trigger plate. By this plan the metal is left in the body just where it is most required to stand the strain of heavy charges. The hinge pin is well forward, which is another advantage. The bolt can be worked by a lever on the side, as shown in the illustration; or by a thumb lever on the break-off. We do not, however, recommend this plan for India in preference to the double-grip; but we believe it to be one of the best snap-actions made.

PENETRATION OF RIFLE BULLETS.

The steel-pointed bullet has always been regarded as the best kind for penetration, but when fired at the skull of an elephant, it has been discovered that the lead separates from the steel point on striking; this is owing to the diameter of the lead being greater than that of the steel point. This cannot be avoided, for if the steel point is not kept smaller than the lead, it will scratch and injure the rifling of the barrel. Hardened lead bullets are far better for this purpose.

12-bore double rifles are made for elephant-shooting, taking a cartridge case which will contain 8 drams of powder, with a hardened spherical or short conical bullet. The weight of these rifles should be 13 to 14 lbs.

Some sportsmen prefer the Henry bullet, 450-bore, weighing 480 grains. This is slightly hardened, and with a charge of 85 grains, we can penetrate six 1-inch elm boards nailed one inch apart, at a distance of 12 yards; but the bullet gets flattened very much (nearly five times its diameter). By hardening the bullet more, a still better penetration may be obtained. We have tried a bullet much lighter, 350 grains, with 4 drams of powder, but the penetration with this is but five boards.

The bullet should not be lighter than 480 grains, and made as hard as possible, composed of two-thirds of old type, and one-third of lead; or, if alloyed with tin or antimony, it will answer quite as well.

We, however, maintain that there is nothing so effective as a large-bore shell for elephant-shooting.

We may remark here that a light hollow Express bullet, 450-gauge, with a charge of 4 drams of powder, will penetrate five 1-inch elm boards at 12 yards.

The Snider Express bullet with 3 drams of powder will penetrate four boards, entering the first board by a hole the diameter of the bullet, and increasing in size gradually to 2 inches.

EXPERIMENTS WITH EXPLOSIVE SHELLS.

Shells for large-bore rifles, such as the "Copper-bottle" and Forsyth swedge shell, have been thoroughly tried on every kind of large game, and their utility in instantly stopping an animal is well known.

The effect of an explosive shell on an animal is much more paralysing than a wound from a solid bullet.

There is an impression among sportsmen that a small quantity of detonating compound, such as is used in the Forsyth shell, would not answer in small-bore shells. To satisfy ourselves on this point, we loaded an ordinary Express bullet by filling up the hollow with the explosive compound and closing the point with wax, firing it with a charge of 3 drams of powder, into the head of a bullock. The bullet penetrated the skull and entered the brain, and, on examination, there appeared merely a small hole in the forehead; but, on opening the head, we found the brain completely destroyed; the shell had burst into small pieces, fracturing the bones. We then tried a 450 Express bullet, weighing 300 grains, in the same manner. The hole in this bullet, being smaller, did not contain so much detonating powder, but the results were about the same with the exception of the penetration; this was greater in consequence of the increased charge, which was 4 drams. The shell exploded more at the back of the head, completely shattering it. The rifle was fired at a distance of 15 and 40 yards respectively, with the same results.

Our next experiment with these explosive shells was at a 3-inch walnut plank. The Snider shell entered by a small hole, and increased in size considerably to the depth of 2 inches, when it exploded, ripping up the far side of the plank. We also tried both the Snider and 450-gauge shells at a bag of damp sawdust. We found all the shells had exploded after passing nearly through.

We have proved satisfactorily that an ordinary conical bullet, having a deep hollow similar to an Express, is quite as effective as the copper-bottle or the Forsyth swedge shell loaded with the same compound, which is sulphuret of antimony and chlorate of potass in equal parts. It can be used in any size bore from 450 to 8. Light, short, conical shells can be used in shot-guns at close quarters with advantage. This shell bullet can easily be reduced to the weight of a solid spherical by increasing the size of the hollow. The advantage of these light shells is that they can be used successively in light guns and rifles, and will be found as effective as large bores with solid bullets.

TO LOAD THE EXPRESS CARTRIDGES.

The first thing to be observed is to carefully wrap the bullet in a prepared paper supplied by the gun-maker, which is exactly the same as that used by the Government for the Martini-Henry. This process must be strictly adhered to if accurate shooting is desired, as the Express bullet is a failure if it is not observed. The most convenient method is as follows :—

Take one of the papers and roll it on the stick to form the case. Place the stick on the paper where marked, keeping the thin end of the stick to the right hand; roll it firmly and level, up to the line marked for gum. Gum it thinly with a camel-hair pencil, and finish rolling it. The paper may then be taken off the stick and left to dry.

When the case is dry put in a bullet at the large end, forcing it through with the rolling stick, leaving only the point uncovered for about one-quarter of an inch; the paper is then cut off, leaving just sufficient to cover the base of the bullet. Lubricate the bullet with Russian tallow, which can be melted in a common glue-pot; when in a liquid state grasp the lead portion of the bullet with the thumb and finger, and dip it up to the top of the paper. The bullets should be first warmed. The next process is to put the powder into the case, then a cardboard wad, then a lubricated thin felt wad, then another cardboard wad, then put in the wrapped and lubricated bullet. A creasing machine is used sometimes to secure the bullet in the case; but this is not absolutely necessary, as the bullet is tolerably firm without being creased.

No. 1 is the ordinary Express bullet, 450-bore, weighing 320 grains. No. 2 is the Westley-Richards form, 260 grains. No. 3 is a light Express, 500-bore, coated with wax instead of paper. A bullet of No. 3 description, weighing 400 grains, can be made to shoot most accurately when properly fitted in a 577 double rifle or a light single sporting Snider, increasing the point-blank range very considerably.

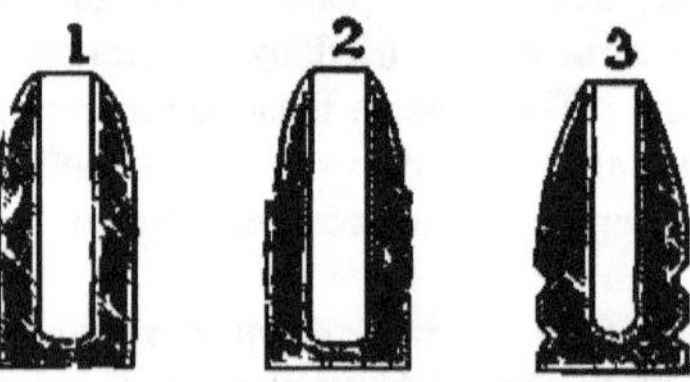

Fig. 55.—Bullets for Express Rifles.

Upwards of 3 drams of powder can be used with little recoil. This improves the weapons for sporting purposes.

EXPLOSIVE SHELLS.

We give an engraving of Forsyth's swedge shell; also the copper bottle shell.

The first shells invented contained a small charge of black powder, which was ignited by an ordinary percussion cap placed on the point of the bullet, and which exploded on striking. The Jacob shell is an improvement, being a copper tube open at one end, containing both detonating and black powder. This shell is found to be rather uncertain in exploding, therefore Forsyth brought out the swedge shell as an improvement. This is cast in two segments, the detonating compound is then put in, the base of the bullet is joined to the other part, and passed through a screw swedge, which, if properly made, makes the bullet appear as one piece. These shells are only adapted for large-bore rifles—16, 12, or 10-bores. The apparatus for making these shells is rather bulky and expensive, consisting of two pairs of moulds and one swedging machine.

Nos. 1 and 2 are the copper bottle, No. 3 is the Forsyth, No. 4 figure represents the Forsyth swedge shell, Nos. 5 and 6 the new segmental shell. No. 5 is cast in a mould by means of a core-peg, having four wings, which divides the mould into four chambers. The segments are then

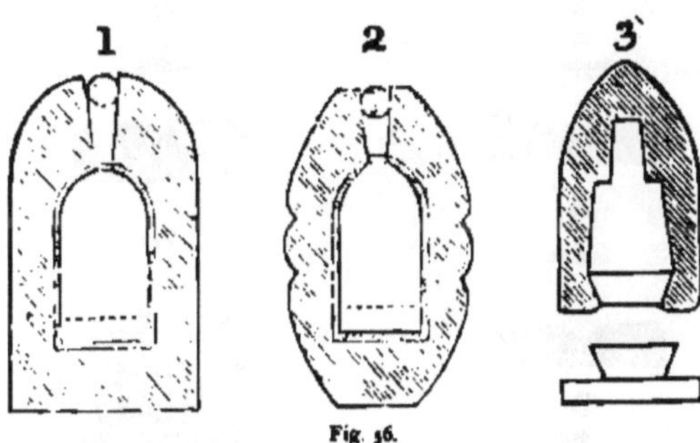

Fig. 56.

tied together, placed on a thin core-peg, put into a larger mould, and a thin jacket of lead cast round them, leaving a small hole at the point. The intention is that the bullet shall fly into pieces on striking, without the use of detonating powder.

The copper bottle shell alluded to above is merely cast in a simple mould, first placing the copper bottle inside, fixing it on a core-peg; this keeps it in the proper position to receive the lead. This peg is withdrawn, and leaves the opening to admit of the detonating powder being put in, the orifice is then closed with wax.

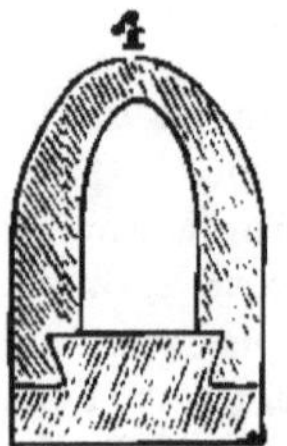

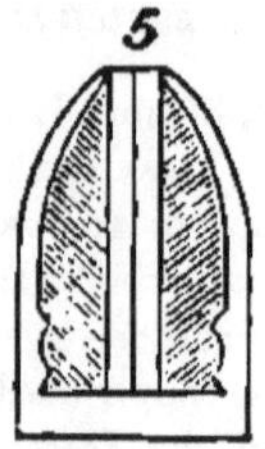

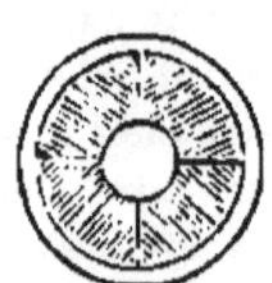

Fig. 37.

Fig. 58.

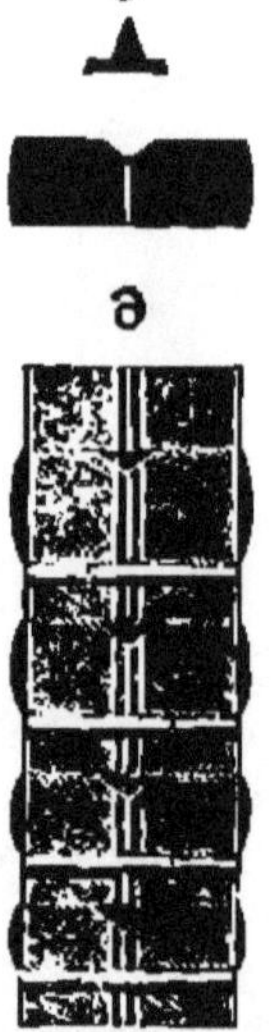

Fig. 59.

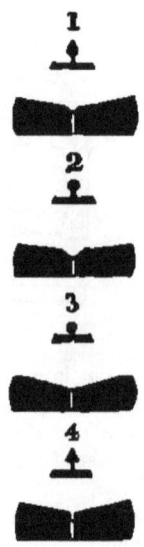

THE EXPLOSIVE COMPOUND FOR SHELLS.

The detonating powder used in shells is not easily exploded. The blow given to the shell on striking will explode it. It can be made, however, very sensitive, and requires care in mixing, which should be done as follows:—Take sulphuret of antimony and chlorate of potass, pounded separately, and mixed carefully in equal parts by weight with a bone knife, on a plate or some smooth surface.

SIGHTS FOR SPORTING RIFLES.

There are various forms of sights used in sporting rifles. We give an illustration of those most generally used.

No. 3 is considered the best for fine shooting. The V is broad, and extends the whole width of the leaf, having a platina line to mark the centre. Sometimes a small slot is preferred, as in No. 4. The muzzle sights are put in lengthways, instead of across the rile as formerly. No. 1, bead sight, is a good form for large-bore rifles; No. 3 is also adapted for rough work, and not easily broken; Nos. 2 and 4 are suitable for Express or target rifles; No. 6 is a leaf sight roughed to prevent reflection. The muzzle sights are frequently inlaid with platina for jungle shooting.

BREECH-LOADING PEA-RIFLES.

We give a representation of a good breech-loading pea-rifle. The bore is 380, and adapted for receiving a brass cartridge containing a charge of 1½ dram of powder. It is snap-action; the hammer and striker are all in one, similar to that of a central-fire revolver. The barrels are made of drilled steel, and octagon in shape. The rifle is light, weighing only 5¼ lbs.; the shooting is very accurate. This is the best kind of pea-rifle made, and the price is moderate. For rooks, rabbits, pea-fowl, or even small deer, it will be found well adapted. The trajectory is very low, and the hollow Express bullet can be used with it. The cartridge-cases are supplied

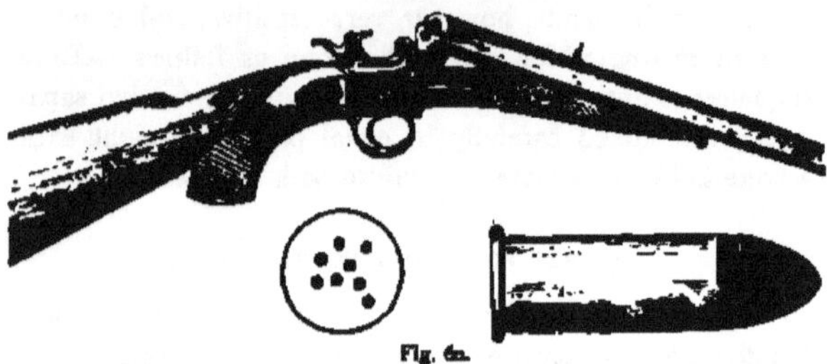

Fig. 61.

with the rifle, empty, and can be filled by the sportsman, which is an advantage when the rifle is required for exportation.

THE MAYNARD SPORTING RIFLE.

This is an American-made rifle, taking a metallic cartridge, which is ignited by the ordinary cap and nipple. There is a small hole in the base of the cartridge to admit the flash. It shoots well at short ranges.

RIFLING FOR SMALL-BORE RIFLES.

The illustration Fig. 62 shows the Henry barrel cut down the middle; it has seven grooves, with one turn in twenty-two inches. The grooves are shallow, with the angles rounded. We also give a representation of the bullet suitable for this rifling, which is of hardened lead and 480 grains. The bullet is coated with paper, and expands in the bore, the base being driven in by the force of the explosion of the large charge of powder, thus filling up the grooves of the rifling entirely, and preventing windage. For lubrication, a wax wad

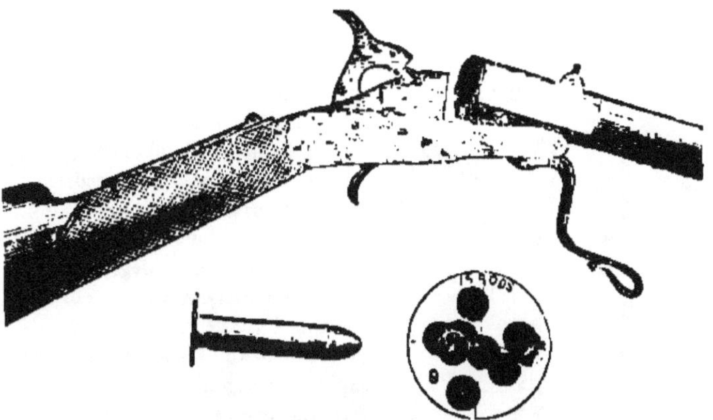

Fig. 61.—The Maynard Sporting Rifle.

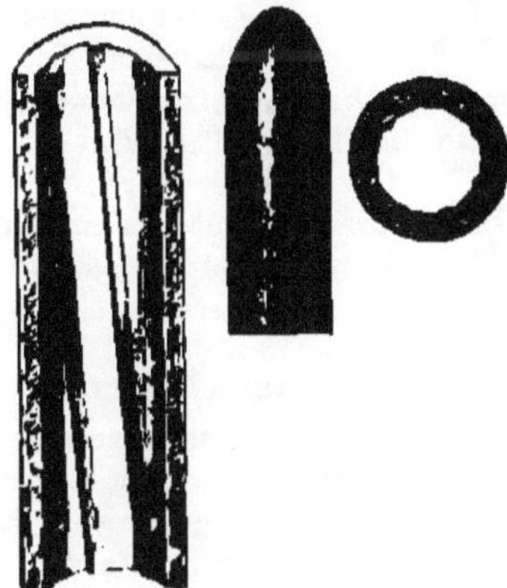

Fig. 62.—The Henry Rifling.

is used over the powder. This is a modification of the Whitworth system, and is found to answer better than any other for small-bore breech-loaders.

Fig. 63 shows a kind of rifling which answers well for small-bore Express rifles. It takes the same bullet as the Henry when rifled with the same degree of spiral, for military purposes. For sporting Express rifles, one turn in twenty-two inches is too rapid. Each gun-maker has his favourite degree of spiral, varying from one turn in twenty-five inches to one in forty inches.

Fig. 63.

We now give a sketch of the Chassepôt rifling (Fig. 64). It has four grooves, with a spiral of one in twenty-one; the bore is smaller than the Henry. The bullet is a trifle shorter, and of a similar shape; the weight is 380 grains. The bullet is larger than the bore of the barrel, and is driven through the rifling, which impairs accuracy.

Fig. 64.

THE SNIDER BULLET AND RIFLING.

The bore of the Snider is 577, or about 25 gun-gauge. There are five grooves, and one turn in four feet. The weight of bullet is 530 grains, it has cannelures to hold the wax for lubrication, and a deep hollow in the base, which is filled up by a clay plug. This plug is driven into the bullet by the force of explosion, and causes it to expand, filling up the grooves of the rifling, and preventing windage. The bullet is made sometimes with a hollow in the point, filled up with wood; but the latest improvement is to leave out the wood plug, and close the hole by compressing the point. This is called the Metford pattern. The motive for making use of the hollow is to lengthen the bullet, whilst retaining the same weight. A long bullet takes the rifling more effectually, and gives better shooting. Fig. 65 represents this bullet and a cross-section of the barrel.

Fig. 65.

STEEL CHAMBERS FOR BREECH-LOADERS.

Fig. 66 represents a steel cartridge case, having a nipple screwed in the base to take the ordinary percussion cap, which is exploded by the needle of a central-fire breech-loader. These cases can be used an indefinite number of times, and are useful when the paper cartridge cannot be obtained readily. The objection to them is the escape of gas through the nipple-hole. We consider these contrivances are now

Fig. 66.

superseded by the brass-drawn "Berdan" case, which is lighter, cheaper, and made for sporting guns of 10 and 12 gauge. We have tried these Berdan shells, and find them answer exceedingly well. They can be recapped any number of times. The shot is kept in by a large tight-fitting felt wad.

Fig. 67 is a cartridge case for a pin breech-loader.

Fig. 68 is a cartridge case for a central-fire breech-loader.

These cases are turned down, or closed upon the wad by a simple machine. There are several descriptions of cartridge machines. We consider Hawksley's patent to be the best for central-fire. The Northcote and Erskine

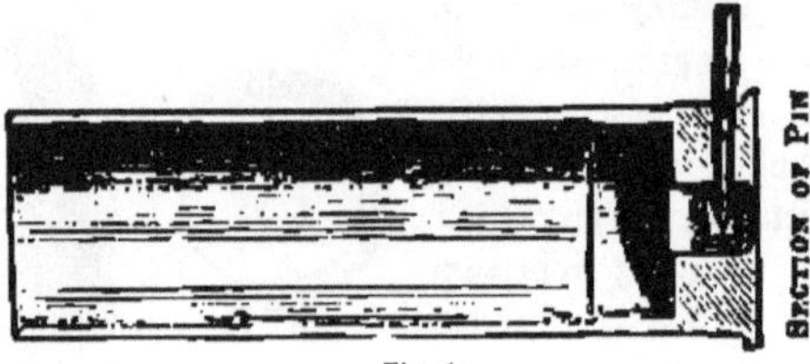

Fig. 67.

machines are useful where large quantities of cases are required to be loaded in a short time; but they are expensive. The Northcote, being a magazine loader, fills the

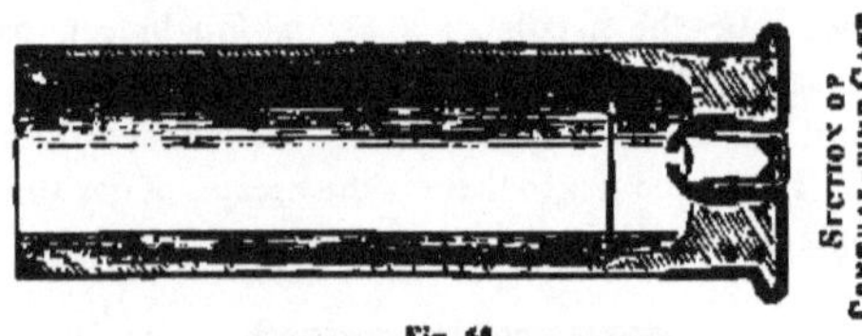

Fig. 68.

cases rapidly; they are best made with a separate closer. The Erskine is a good machine, but rather bulky.

HOW TO SELECT OR ORDER A GUN.

The first consideration is to ascertain the weight you can carry comfortably. Have a straight stock if you can. On putting the gun to your shoulder, catch the centre of the rib with your eye quickly; if you cannot do this the gun is not crooked enough in the stock for you. This is a point of great importance in the selection of a gun; because if you are not properly fitted, and the gun is not handy, you cannot possibly make good shooting. Having satisfied yourself of the weight, length, and bend of stock which suit you, take the measure of the gun as follows, viz. :—From the front or right trigger to the centre of the heel plate; lay the gun on a table and draw a straight line the whole length of the barrels along the rib to the end of stock, as shown in the sketch.

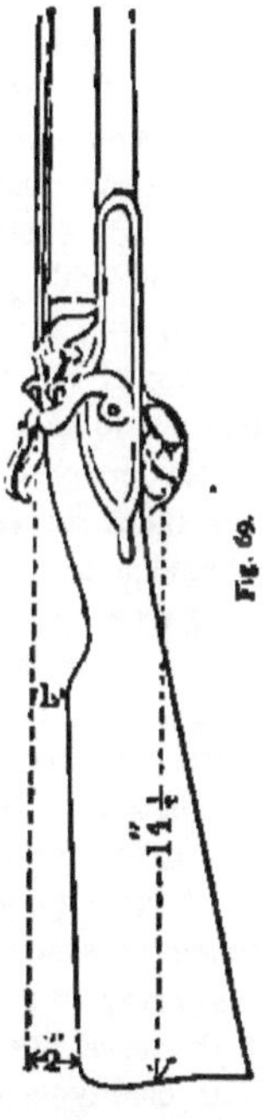

Another plan is to place a straightedge along the rib, extending backwards to the heel-plate. Then measure the distance between the stock and the straightedge. A tracing of the gun will do, provided that it is taken on stiff paper, and at least half the length of barrel to be included. Care must be taken to give the exact position of the right trigger. The sketch we give is of a gun of the ordinary English length and bend, and would suit sportsmen of the average height. The Americans use guns with much more crook in the stock, viz., 2¾ in. to 3 in.

Fig. 68.

PRICES OF BREECH-LOADING SPORTING GUNS AND RIFLES.

The first-class London makers charge for best double rifles, with cases complete, from £65 to £84 each. The price for shot guns is from £50 to £55 each. Some second-class makers charge for double breech-loading rifles, complete in cases, £57, and guns £45. They may be purchased at a much lower figure from makers of less note. Our best price for double rifles, highly finished, with cases complete, is £45, and shot or ball guns £35. Our special

Indian rifle, best material and workmanship, without ornamentation, is £35, without case. Canvas cases complete, £4; best solid leather, £5; shell moulds extra. The gun to match would be £25, without case. We charge the same for Express rifles as for spherical ball. Double rifles of 577-bore taking the Snider ammunition are made from £20 to £25, without cases. There are a great number of good second-class guns made at £20. Single rifles on the Lefaucheux plan, £16 to £20. Sporting Martini rifles, £10 10s. Sporting Sniders, made to shoot accurately, from £5; if with pistol grip and superior finish, £6 10s. to £8.

Any of the above guns delivered in Calcutta or Bombay, would cost £5 more.

Sniders or Martinis, £1 10s. or £2. Cartridge cases, per thousand (blue), £2. Green, gas-tight, £2 10s. Metal-lined, £3. Snider cases, £2 10s.; with the addition of £1 per thousand delivered in India.

The best and most expeditious route is by the Peninsular and Oriental Steamship Company. Guns, rifles, and cartridges may be obtained from Messrs. Grindlay and Co., Calcutta and Bombay, at about the same prices as are charged at the works, with the addition of freight, &c.

A large quantity of guns are sent annually to the United States of America from Birmingham, at moderate prices. Not many first-quality guns are exported, on account of the high rate of duty, which is more than one-third of the cost. The quality usually supplied is from £10 to £15, without cases. A very large number of muzzle-loaders are still sent at much lower prices. A good serviceable gun, suitable for a settler, may be obtained for £7.

Military rifles vary in price according to the demand, and on being machine or hand made. Good Sniders, machine made, interchangeable, can be obtained in quantities at

£2 10s. The Remington and the Berdan, a little cheaper. The Westley-Richards, about £2 10s. The Soper and Martini rifles, when machine-made, ought not to cost more without the royalty. The Chassepôt, which is made partly by hand and partly by machinery, costs £2 15s. with the sword bayonet. The Snider made by hand, with triangle bayonet, £2 5s.; with sword bayonet, £2 8s. Good arms can be supplied at the above prices in large quantities, in quiet times, but during the Franco-Prussian war the prices were double.

Loaded Boxer cartridges, in large quantities, can be supplied from £4 per thousand. The Chassepôt cartridge, about £3 per thousand.

THE PRICE OF REVOLVERS.

Adams' and Colt's are the most expensive revolvers. They are sold at prices varying from £5 10s. to £7 10s., in cases complete. Tranter's, Braendlin's, and Thomas's are about £5 each. Webley's constabulary revolver, and Greener's central-fire, £4 10s. Smith and Weston's, £3 10s. Revolvers for pin cartridge, seven, nine, and twelve millimetre, from £1 5s. to £2, without cases. Double-barrel pistols for the Boxer cartridge, with locks to screw off, £8 each; second quality, £6. Cartridges for these revolvers cost from £2 5s. to £2 15s. per thousand.

THE GUN-BARREL PROOF ACT, 1868.

We give illustrations of the proof-marks used by each company, viz., the London and Birmingham. There is an erroneous impression abroad that gun-barrels being stamped with the London marks must be London-made. This is not the case; many gun-makers send their barrels to London to be proved, because guns so marked find a readier sale among dealers.

Gun-barrels are proved by the same scale both in London and Birmingham, and with the same strength of powder adopted by the Government for small arms. We introduce a tabular list of the proof charges.

Barrels with foreign proof-marks are exempted, except in case of being marked as of English manufacture.

Fig. 70.

London Marks. Birmingham Marks.

Foreign barrels converted in England are deemed unproved, but if they are only converted and immediately exported they are exempt from penalties, but due notice must in all such cases be given to the proof-masters. Old muzzle-loaders intended for conversion must be subjected to provisional and definitive proof, if of English manufacture.

THE GUN-BARREL PROOF ACT, 1868.

SCALE FOR PROOF OF RIFLED SMALL ARMS OF EVERY DESCRIPTION.

Number of Gauge.	Diameter of Bore.	Bullet for Proof.				Charge of Powder for						Service Charge.			
		Diameter.	Length.	Ratio of Length to Diameter.	Weight.	First Proof.			Second Proof.			Powder.			Ball.
	Inches	inches	inches		grains	grs	oz.	drs.	grs.	oz.	drs.	grs	oz.	drs.	grs.
1	1.669	1.649	2.474	1.500	15186.0	3417	7	13	2278	5	3¼	1139	2	9¾	11390
	1.500	1.480	2.220	1.500	10977.0	2470	5	10½	1646	3	12¼	823	1	14	8233
2	1.325	1.305	1.958	1.500	7527.1	1694	3	14	1129	2	9¼	565	1	4¾	5645
	1.250	1.230	1.845	1.500	6300.9	1418	3	3¾	945	2	2½	473	1	1¼	4726
3	1.157	1.137	1.700	1.500	4978.5	1120	2	9	747	1	11½	373	0	13½	3734
4	1.052	1.032	1.548	1.500	3721.6	837	1	14½	558	1	4½	279	0	10¼	2791
	1.000	.980	1.470	1.500	3186.9	717	1	10¼	478	1	1½	239	0	8¾	2390
5	.976	.956	1.434	1.500	2958.4	666	1	8¼	444	1	0¼	222	0	8	2219
6	.919	.899	1.349	1.500	2461.1	554	1	4¼	369	0	13½	185	0	6¾	1846
	.900	.880	1.320	1.500	2307.5	519	1	3	346	0	12¾	173	0	6¼	1731
7	.873	.853	1.280	1.500	2102.3	473	1	1¼	315	0	11½	158	0	5¾	1577
	.850	.830	1.245	1.500	1936.1	436	1	0	290	0	10½	145	0	5¼	1452
8	.835	.815	1.223	1.500	1833.7	413	0	15	275	0	10	138	0	5	1375
9	.803	.783	1.175	1.500	1626.1	366	0	13½	244	0	9	122	0	4½	1220
	.800	.780	1.170	1.500	1606.8	362	0	13¼	241	0	8¾	121	0	4½	1205
10	.775	.755	1.132	1.500	1456.6	328	0	12	219	0	8	109	0	4	1092
	.770	.750	1.126	1.501	1429.7	322	0	11¾	215	0	7¾	107	0	4	1072
	.760	.740	1.112	1.503	1374.6	310	0	11¼	207	0	7½	103	0	3¾	1031
11	.751	.731	1.100	1.505	1326.9	302	0	11	200	0	7¼	100	0	3¾	995
	.750	.730	1.099	1.505	1322.0	300	0	11	200	0	7¼	100	0	3¾	992
	.740	.720	1.083	1.507	1269.7	289	0	10½	193	0	7	96	0	3½	952
	.730	.710	1.071	1.509	1218.7	279	0	10¼	186	0	6¾	93	0	3½	914
12	.729	.709	1.070	1.509	1214.2	278	0	10¼	185	0	6¾	93	0	3½	911
	.720	.700	1.058	1.511	1170.2	269	0	9¾	180	0	6½	90	0	3¼	878
13	.710	.690	1.045	1.514	1123.1	260	0	9½	173	0	6¼	87	0	3¼	842
	.700	.680	1.032	1.518	1077.2	252	0	9¼	168	0	6¼	84	0	3	808
14	.693	.673	1.024	1.521	1047.0	246	0	9	164	0	6	82	0	3	785
	.690	.670	1.020	1.523	1033.6	244	0	9	162	0	6	81	0	3	775
	.680	.660	1.003	1.529	992.1	236	0	8¾	158	0	5¾	79	0	3	744
15	.677	.657	1.006	1.531	980.2	233	0	8½	156	0	5¾	78	0	2¾	735
	.670	.650	.999	1.537	952.8	230	0	8½	154	0	5½	77	0	2¾	715
16	.662	.642	.992	1.545	923.0	226	0	8¼	151	0	5½	75	0	2¾	692
	.660	.640	.990	1.547	915.4	225	0	8¼	150	0	5½	75	0	2¾	687
	.650	.630	.982	1.559	879.8	220	0	8	147	0	5¼	73	0	2¾	660
17	.649	.629	.981	1.560	876.1	220	0	8	146	0	5¼	73	0	2¾	657
	.640	.620	.977	1.575	847.8	217	0	8	145	0	5¼	72	0	2½	636
18	.637	.617	.975	1.581	837.9	216	0	8	144	0	5¼	72	0	2½	628
	.630	.610	.974	1.596	818.1	214	0	7¾	143	0	5¼	71	0	2½	614
19	.626	.606	.973	1.606	806.6	213	0	7¾	142	0	5¼	71	0	2½	605
	.620	.600	.974	1.623	791.5	212	0	7¾	141	0	5¼	71	0	2½	594
20	.615	.595	.975	1.638	779.2	211	0	7¾	141	0	5¼	70	0	2½	584
	.610	.590	.978	1.657	768.5	211	0	7¾	141	0	5¼	70	0	2½	576
21	.605	.585	.980	1.675	757.1	210	0	7¾	140	0	5	70	0	2½	568
	.600	.580	.985	1.698	748.0	209	0	7½	140	0	5	70	0	2½	561
22	.596	.576	.988	1.715	739.9	209	0	7½	139	0	5	70	0	2½	555
	.590	.570	.995	1.746	729.7	208	0	7½	139	0	5	69	0	2½	547
23	.587	.567	.999	1.762	725.0	207	0	7½	138	0	5	69	0	2½	544

Number of Gauge.	Diameter of Bore.	Bullet for Proof.				Charge of Powder for						Service Charge.			
		Diameter.	Length.	Ratio of Length to Diameter.	Weight.	First Proof.			Second Proof.			Powder.			Ball.
	inches	inches	inches		grains.	grs.	oz.	drs.	grs.	oz.	drs.	grs.	oz.	drs.	grs.
	.580	.560	1.012	1.808	716.4	205	0	7½	137	0	5	68	0	2½	537
24	.579	.559	1.015	1.816	716.0	205	0	7½	137	0	5	68	0	2½	537
Regulation Bore	.577	.557	1.021	1.833	715.0	205	0	7½	137	0	5	68	0	2½	536
25	.571	.551	1.043	1.893	715.0	205	0	7½	137	0	5	68	0	2½	536
	.570	.550	1.047	1.904	715.0	205	0	7½	137	0	5	68	0	2½	536
26	.563	.543	1.074	1.978	715.0	205	0	7½	137	0	5	68	0	2½	536
	.560	.540	1.086	2.012	715.0	205	0	7½	137	0	5	68	0	2½	536
27	.556	.536	1.103	2.057	715.0	205	0	7½	137	0	5	68	0	2½	536
28	.550	.530	1.128	2.128	715.0	205	0	7½	137	0	5	68	0	2½	536
29	.543	.523	1.158	2.214	715.0	205	0	7½	137	0	5	68	0	2½	536
	.540	.520	1.171	2.253	715.0	205	0	7½	137	0	5	68	0	2½	536
30	.537	.517	1.185	2.291	715.0	205	0	7½	137	0	5	68	0	2½	536
31	.531	.511	1.213	2.374	715.0	205	0	7½	137	0	5	68	0	2½	536
	.530	.510	1.218	2.388	715.0	205	0	7½	137	0	5	68	0	2½	536
32	.526	.506	1.237	2.445	715.0	205	0	7½	137	0	5	68	0	2½	536
33	.520	.500	1.267	2.534	715.0	205	0	7½	137	0	5	68	0	2½	536
34	.515	.495	1.293	2.612	715.0	205	0	7½	137	0	5	68	0	2½	536
35	.510	.490	1.319	2.692	715.0	205	0	7½	137	0	5	68	0	2½	536
36	.506	.486	1.341	2.759	715.0	205	0	7½	137	0	5	68	0	2½	536
37	.501	.481	1.369	2.846	715.0	205	0	7½	137	0	5	68	0	2½	536
	.500	.480	1.375	2.864	715.0	205	0	7½	137	0	5	68	0	2½	536
38	.497	.477	1.392	2.918	715.0	205	0	7½	137	0	5	68	0	2½	536
39	.492	.472	1.422	3.012	715.0	205	0	7½	137	0	5	68	0	2½	536
	.490	.470	1.434	3.051	715.0	205	0	7½	137	0	5	68	0	2½	536
40	.488	.468	1.446	3.090	715.0	205	0	7½	137	0	5	68	0	2½	536
41	.484	.464	1.472	3.171	715.0	205	0	7½	137	0	5	68	0	2½	536
42	.480	.460	1.497	3.254	715.0	205	0	7½	137	0	5	68	0	2½	536
43	.476	.456	1.523	3.341	715.0	205	0	7½	137	0	5	68	0	2½	536
44	.473	.453	1.544	3.407	715.0	205	0	7½	137	0	5	68	0	2½	536
	.470	.450	1.564	3.476	715.0	205	0	7½	137	0	5	68	0	2½	536
45	.469	.449	1.571	3.499	715.0	205	0	7½	137	0	5	68	0	2½	536
46	.466	.446	1.592	3.570	715.0	205	0	7½	137	0	5	68	0	2½	536
47	.463	.443	1.614	3.643	715.0	205	0	7½	137	0	5	68	0	2½	536
	.460	.441	1.636	3.718	715.0	205	0	7½	137	0	5	68	0	2½	536
48	.459	.439	1.644	3.744	715.0	205	0	7½	137	0	5	68	0	2½	536
49	.456	.436	1.666	3.822	715.0	205	0	7½	137	0	5	68	0	2½	536
50	.453	.433	1.689	3.902	715.0	205	0	7½	137	0	5	68	0	2½	536
Small Bore	.451	.431	1.705	3.956	715.0	205	0	7½	137	0	5	68	0	2½	536
51.05	.450	.430	1.711	3.978	714.1	205	0	7½	137	0	5	68	0	2½	536
54.61	.440	.420	1.757	4.183	697.6	203	0	7½	135	0	5	68	0	2½	523
58.50	.430	.410	1.781	4.344	673.8	198	0	7¼	132	0	4¾	66	0	2⅖	505
62.78	.420	.400	1.786	4.465	645.1	191	0	7	127	0	4⅔	64	0	2⅓	484
67.47	.410	.390	1.781	4.567	611.5	183	0	6¾	122	0	4½	61	0	2¼	459
72.68	.400	.380	1.769	4.654	576.6	174	0	6⅓	116	0	4¼	58	0	2	432
78.41	.390	.370	1.749	4.728	540.5	165	0	6	110	0	4	55	0	2	405
84.77	.380	.360	1.724	4.790	504.4	155	0	5⅔	103	0	3¾	52	0	2	378
91.83	.370	.350	1.695	4.843	468.7	145	0	5⅓	97	0	3½	48	0	1¾	352
99.70	.360	.340	1.663	4.890	434.0	137	0	5	91	0	3⅓	45	0	1⅔	325
108.49	.350	.330	1.627	4.930	400.0	126	0	4⅝	84	0	3	42	0	1½	300
118.35	.340	.320	1.587	4.960	356.8	117	0	4¼	78	0	2¾	39	0	1⅖	275
129.43	.330	.310	1.543	4.978	334.7	108	0	4	72	0	2⅔	36	0	1⅓	251
141.95	.320	.300	1.497	4.990	304.1	99	0	3⅝	66	0	2⅖	33	0	1¼	228
156.14	.310	.290	1.449	4.997	275.1	90	0	3¼	60	0	2¼	30	0	1	206
172.28	.300	.280	1.400	5.000	247.8	82	0	3	55	0	2	27	0	1	186

BREECH-LOADING REVOLVERS.

We now intend giving a description of all the best kinds of breech-loading pistols and revolvers. We find that the pistol was introduced about the reign of Henry VIII., and was known as the "petronel." The name is derived from the city of Pistola Perugea, in the Romagna. Many specimens of the ancient revolver are to be found in the Museum of Paris, and elsewhere. The inefficiency of these revolvers was not so much owing to the fault of the mechanism as to the mode of ignition; the flint-and-steel or match lock was ill adapted for this kind of weapon. The revolver was again revived when percussion caps came into use. Colonel Colt was the first to produce a truly reliable

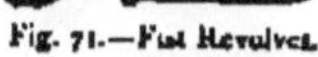

Fig. 71.—Fist Revolver.

Fig. 72.—The Pocket Revolver.

revolver, which was adopted universally by the American, and afterwards by the English Government.

M. Lefaucheux has the credit of bringing out the first breech-loading revolver. The first specimens of the Lefaucheux pin cartridge case were made of paper with metal bases, similar to the gun cartridges. The next improvement was the solid metal case, with a bullet fixed in it, making it waterproof. We give several illustrations relating to this kind of revolver. Fig. 71 is called a Fist Revolver.

Fig. 72 has seven chambers with a barrel attached, and is double-actioned—can be cocked by the thumb or by pulling the trigger. We also give a representation of a ten-shot revolver, which is generally made twelve-millimetre; this is the largest size made, five-millimetre being the smallest.

Seven and nine are most in demand as pocket revolvers, the twelve being generally chosen for use on horseback.

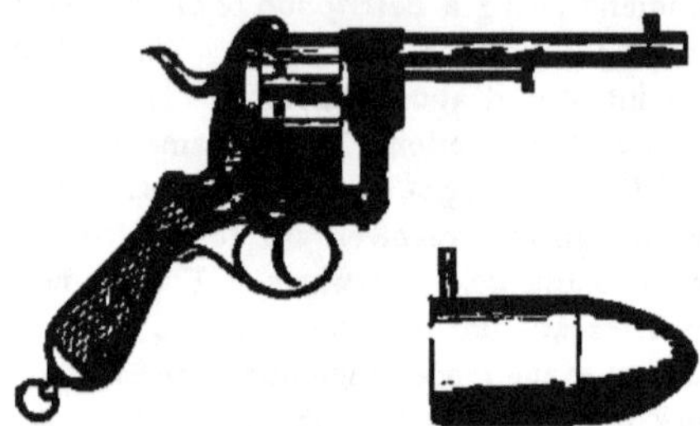

Fig. 73.—The ten-shot Revolver.

Seven chambers are considered sufficient for all purposes, and are tolerably handy; the ten-chamber arrangement being

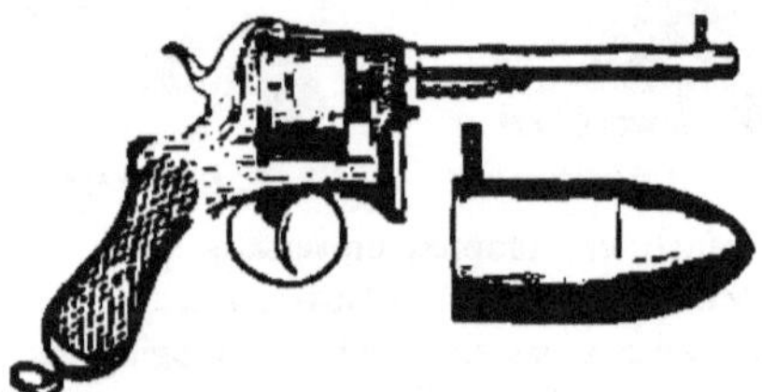

Fig. 74.—The seven-shot Revolver.

very bulky and not much in favour. These revolvers are all made of a cheap kind. The best English makers prefer constructing them on the central-fire plan. The great defect in these pin revolvers is that ten to fifteen per cent. missfire. With a well-made central-fire, missfires are of rare occurrence.

SMITH AND WESSON'S AMERICAN REVOLVER.

This is a well-made revolver; it is of small size, and can be comfortably carried in the pocket. The cartridge contains the ignition in the rim, but is liable to missfire. When first introduced it was considered an improvement on the

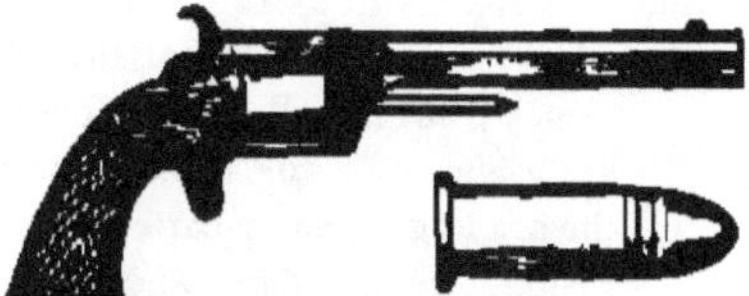

Fig. 75.—The Smith and Wesson Revolver.

pin-fire; but the central-fire is an improvement on that.

CENTRAL-FIRE REVOLVER, 450-BORE.

This is a really useful weapon. The bore is large and very effective; takes a good charge of powder and a conical

Fig. 76.—Central-fire Revolver, 450.

bullet. The range is considerable, and the penetration good.

THE IRISH CONSTABULARY REVOLVER.

The bore of this is 442, with a short barrel which makes it light and handy, with the advantage of a large bore; but it is only good for close quarters.

There is also a revolver made to take the Boxer cartridge case, 577, and a spherical ball, twenty-four to the pound. This is decidedly a most formidable weapon. It has six chambers, and weighs only 2 lb. 14 oz.

Although great improvements have been made in breech-loading revolvers, we do not consider that they can be regarded even now as reliable weapons in cases of great emergency, when human life is dependent on their efficiency; but for all general purposes they are useful. To the subject matter under the heading "Double-barrel Breech-loading Pistols," we refer those who are in search of a good and reliable weapon, for war purposes or for hunting dangerous game—as there can be no doubt that two certain shots are better than six uncertain ones. The revolvers most in demand are Adams' new patent double-action central-fire, and Colt's new patent. We here introduce an account of a competitive trial at the Royal Arsenal, Woolwich, Colt *versus* Adams; a drawing of the weapon, and a target pattern made by Mr. Adams' revolver.

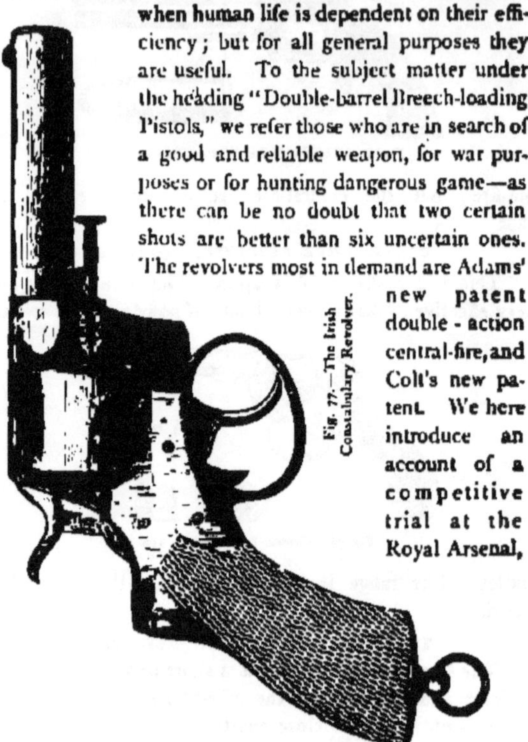

Fig. 77.—The Irish Constabulary Revolver.

ADAMS' PATENT CENTRAL-FIRE BREECH-LOADING REVOLVER.

This revolver has six chambers. It can be cocked for deliberate aim, or discharged in rapid succession by merely pulling the trigger.

INSTRUCTIONS.

To Load.—Half cock the revolver, raise the shield which

Fig. 78.—Adams' Revolver.

partially covers the chambers at the rear end of the cylinder, and insert the cartridges into the chambers.

To Release the Ejector.—Take the ejector (which is held in its proper position by a slight spring) by the cross-head, as it lies at its extreme distance close to the underside of the barrel, and with the forefinger force it outwards at a right angle; it can then be held by the finger and thumb and pushed into the chambers.

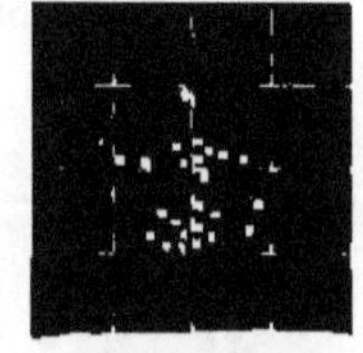

Fig. 79.—Target of Adams' Revolver.

To Lock the Cylinder.—Release the ejector, and push it into one of the chambers (turning, at the same time, the cross-head to the underside of the barrel) until the second stop in the ejector is caught by the spring bolt.

To Unload.—With the revolver still at half-cock, raise the shield, turn the muzzle upwards, rotate the cylinder slowly, and the cartridges will drop out. If they stick, push them out with the ejector.

To Release the Cylinder.—With the ejector at its extreme distance, and the cross-head lying close to the underside of the barrel, push in with the second finger of the right hand the projecting bolt, which is in the underside of the pistol, and close to the head of the cylinder rod; then with the forefinger withdraw the cylinder rod as far as the cross-head of the ejector; the cylinder can now be taken out.

After firing the revolver, the ejector is required to knock out the empty cartridge cases.

Well oil the revolver after use, and always clean it with oil.

The cartridges have been specially designed for this revolver by Colonel Boxer, R.A., Chief Superintendent of the Royal Laboratories, Woolwich.

COMPETITION TRIAL OF THE COLT AND THE ADAMS BREECH-LOADING REVOLVERS AT THE ROYAL ARSENAL, WOOLWICH.

The following is a digest of the results of the trials, and we reproduce them in this simple form, believing that they will be more intelligible to the majority of readers than in the original text:—

Particulars of Pistols and Ammunition.

Name.	Weight of Pistol.	Calibre.	Length of Barrel.	Charge.	Lubrication.	Weight of Bullet.
Adams	2 lb. 6[illegible] oz.	·450	6 in.	13 grs.	Beeswax in two cannelures.	225 grs.
Colt ...	2 lb. 10[illegible] oz.	·440	8 in.	18 grs.	Wad of bees-wax.	210 grs.

Each arm was fired by its own representative.

Rapidity of Fire.—In this trial twenty-four shots had to be fired without any attempt at accuracy, the time commencing from the loading of the first shot and ending with the ejection of the last cartridge case.—Adams, 1 min. 30 sec.; Colt, 2 min.

A second experiment was also made with ammunition that had been purchased from the respective competitors in the ordinary way of business. In this case the times were —Adams, 1 min. 35 sec.; Colt, 2 min. 20 sec.

Accuracy.—The trials for accuracy were divided into two classes, viz., those for accuracy only, irrespective of time, and for accuracy combined with rapidity.

Penetration at Thirty Yards.

Adams.				Colt.			
1st shot penetrated	...	3·25	boards.	1st shot penetrated	...	4·00	boards.
2nd	,, ...	3·25	,,	2nd	,, ...	3·00	,,
3rd	,, ...	4·25	,,	3rd	,, ..	3·75	,,
4th	,, ...	3·25	,,	4th	,, ...	2·25	,,
5th	,, ...	3·25	,,	5th	,, ...	3·25	,,
6th	,, ...	3·75	,,	6th	,, ...	3·25	,,
Total ...	...	21·00	,,	Total ...	...	19·25	,,
Mean of six shots	...	3·50	,,	Mean of six shots	...	3·25	,,

Penetration at Sixty Yards.

Adams.				Colt.			
1st shot penetrated	...	3·25	boards.	1st shot penetrated	...	3·25	boards.
2nd	,, ...	4·25	,,	2nd	,, ...	3·25	,,
3rd	,, ...	3·25	,,	3rd	,, ...	3·25	,,
4th	,, ...	4·25	,,	4th	,, ...	2·75	,,
5th	,, ...	4·25	,,	5th	,, ...	2·00	,,
6th	,, ...	3·00	,,	6th	,, ...	1·50	,,
Total ...	...	22·25	,,	Total ...	...	16·00	,,
Mean of six shots	...	3·70	,,	Mean of six shots	...	2·66	,,

From these results, then, we find that the two pistols bear the following comparison:—

	Adams.	Colt.
1. Rapidity only (average)	1 min. 32 sec. ...	2 min. 10 sec.
2. Rapidity and accuracy (ditto) ...	2 ,, 18 ,, ...	2 ,, 55 ,,
3. Accuracy (ditto)...	5·92 in. ...	8·29 in.
4. Penetration (ditto)	3·60 boards... ...	2·95 boards.

Reducing these figures to percentage of superiority, we have the Adams superior on each of the points as follows:—

	Adams superior per cent.
1. Rapidity only	41
2. Ditto when combined with accuracy	26
3. Accuracy of shouting (deliberate and rapid fire) ...	40
4. Penetration	22

—*From the Engineer, July* 16, 1869.

COLT'S NEW PATENT METALLIC CENTRAL-FIRE CARTRIDGE REVOLVERS.

While doing away with nipples and the necessity of capping, the manufacturers have retained all the best qualities of the original Colt's revolver. In length of range, accuracy of fire, force of penetration, simplicity of construction, and durability, they equal all other metallic cartridge revolvers.

By means of a simple loading apparatus used in connection with the pistol, the metallic shells can be easily re-loaded and fired as often as may be required.

An interchangeable nipple cylinder can likewise be supplied with these pistols, which can then be loaded with loose powder and ball, a matter of so much importance in countries where made-up ammunition is not easily procurable.

The shells do not expand and stick in the chambers of the cylinder through the explosion of the cartridges,

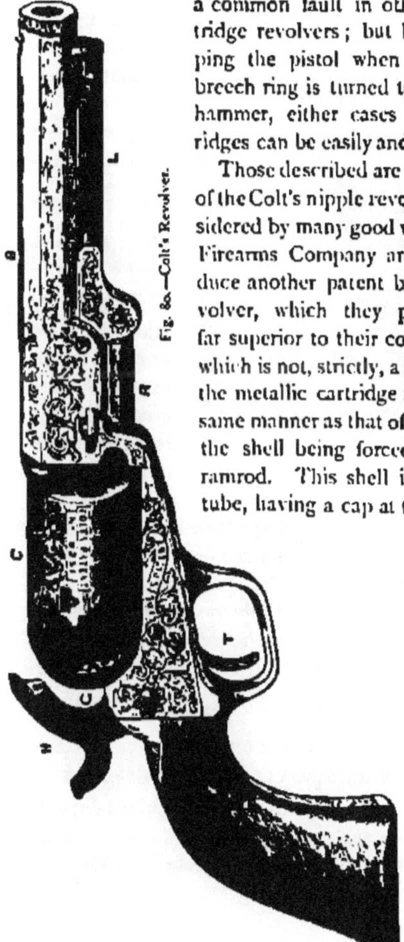
Fig. 80.—Colt's Revolver.

a common fault in other metallic cartridge revolvers; but by simply snapping the pistol when the boss of the breech ring is turned to the left of the hammer, either cases or loaded cartridges can be easily and rapidly ejected.

Those described are only conversions of the Colt's nipple revolver, but are considered by many good weapons. Colt's Firearms Company are about to introduce another patent breech-loading revolver, which they promise will be far superior to their converted weapon, which is not, strictly, a breech-loader, as the metallic cartridge is inserted in the same manner as that of a muzzle-loader, the shell being forced down with the ramrod. This shell is simply a taper tube, having a cap at the base. These cases can be reloaded many times. We give a drawing of the original Colt's revolver—which has much the same appearance as the converted pistol; also one with a movable butt, which makes

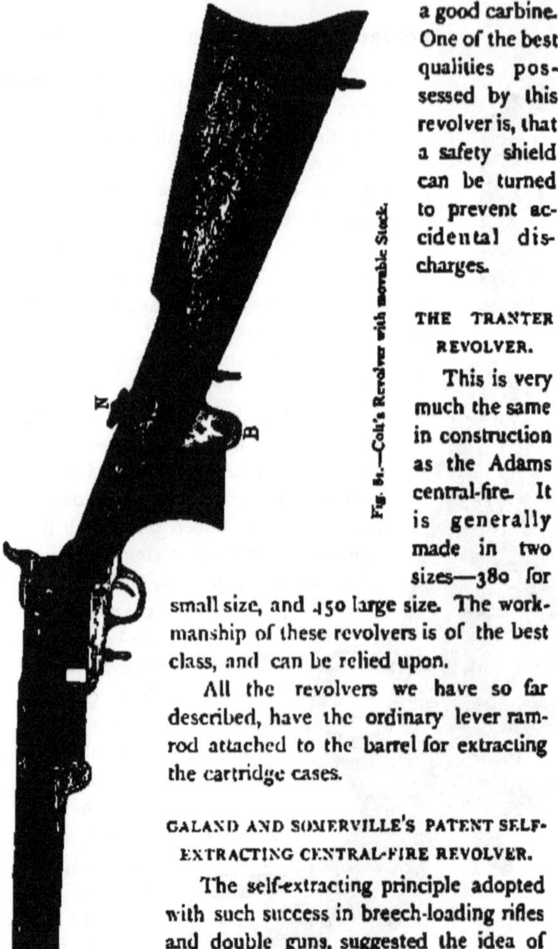

Fig. 81.—Colt's Revolver with movable Stock.

a good carbine. One of the best qualities possessed by this revolver is, that a safety shield can be turned to prevent accidental discharges.

THE TRANTER REVOLVER.

This is very much the same in construction as the Adams central-fire. It is generally made in two sizes—380 for small size, and 450 large size. The workmanship of these revolvers is of the best class, and can be relied upon.

All the revolvers we have so far described, have the ordinary lever ramrod attached to the barrel for extracting the cartridge cases.

GALAND AND SOMERVILLE'S PATENT SELF-EXTRACTING CENTRAL-FIRE REVOLVER.

The self-extracting principle adopted with such success in breech-loading rifles and double guns, suggested the idea of doing away with the difficulty of extraction in central-fire revolvers. The Galand-Somerville is the original revolver by the aid of which the problem was satisfactorily solved. The manipulation is very simple. By depressing the lever, the barrel and chamber move forward, and consequently extract all the empty cases at one time. It can then be reloaded, the lever returned, and it is ready again for firing. This revolver is made by machinery, and the workmanship

Fig. 82.—The Galand and Somerville Patent Self-extractor Revolver.

is very good. It has six chambers, and is double-action; the large size is 450, and the small 380.

THOMAS'S PATENT REVOLVER.

This revolver is constructed on an improved principle.

Fig. 83.—Thomas's Patent Revolver.

The empty cartridge cases are extracted simultaneously by turning over the barrel and drawing it forward about one inch; the empty cases drop out as represented in the illustration (p. 159). It has a solid body; the centre rod being enclosed in a tube prevents its fouling from the escape of gas, a defect found to exist in many other revolvers. It is said that this revolver can be fired 500 times without cleaning. For use on horseback these weapons would be invaluable. They are made in three sizes—450, 380, and 320. Double-action, on the central-fire plan, and made with five chambers.

THE DOUBLE-BARREL BREECH-LOADING PISTOL.

Many sportsmen detest revolvers of any kind, and prefer a good double-barrel pistol. We have succeeded in making a very handy double pistol 577-bore, to take the Snider cartridge case, with spherical ball, twenty-three to the pound.

Fig. 84.—The Double-grip Pistol.

A pistol weighing $3\frac{1}{4}$ lbs. will take a charge of $1\frac{1}{2}$ drams of powder, and give great penetration and smashing power. The advantage possessed by these pistols is, that they take a very large bullet, which, being spherical, gives a tremendous shock to the system of an animal. By increasing the weight of the pistol, a larger charge of powder may be used. It would then answer the purpose of a second rifle for close quarters. Fig. 84 represents this pistol; it is made on the Lefaucheux principle, either with the lever over the guard as in a gun, or snap action—that is, with a lever at the side of the lock. The latter plan is very quick, and

Fig. 85.—Snap-action Pistol.

admits of very easy loading; we prefer it to the double grip action.

These pistols are made with detached locks, similar to those of a gun, and are not likely to get out of order. They are also made with locks similar to officers' pistols (*see* Fig. 85).

There is no doubt that shells will be used in pistols and

Fig. 86.

revolvers, against dangerous game, for the future. The sharp conical revolver bullet will not give sufficient shock to stop an animal. That which is wanted is an expanding bullet, with a hollow point, like an Express or explosive shell. A double-barrelled pistol, 577-bore, loaded with a shell, would be found a most deadly weapon at close quarters. The following account of the experiments with the Pertuiset shell will be convincing.

THE PERTUISET SHELL.

The compound made use of appears to be the invention of a French chemist. We have tried it in explosive shells, and on dead animals, and find that its effects are wonderful. The shell enters the skull by a small hole, leaving the bone fractured in front like a star, and explodes in the brain, scattering it in all directions, completely destroying the whole structure, which must cause instant death in any animal to whose brain one of these shells can enter. We give the *Times* account of the trial of this powder on horses in London, which report we believe to be truthful, judging from what we have witnessed ourselves of this explosive material:—

The Pertuiset powder was first employed for small bullets only, and its adoption by Russia led to the Congress at St. Petersburg, where the principal military Powers of Europe decided not to use it in small-arm ammunition against men. The United States declined to hamper themselves by any such agreement. The inventor claims for his powder absolute safety in use; but we should like to see it well tried before handling it heedlessly. If it be true that the Prussians are about to use it in projectiles fired from the Gatling, its moral effect, when employed against villages, houses, and walls, must be greater than that of the mitrailleuse.

The experiments yesterday, in London, were made for the information of Surgeon-Major Wyatt, and were conducted in a shed attached to the yard of Messrs. Winkley and Shaw, horse slaughterers, at 35, Green Street, Blackfriars Road. Mr. Adams was present with one of his revolvers, and fired all the shots. A small group of officers and others assembled at two o'clock in the afternoon. Mr. Shaw, a partner in the firm, made all the arrangements.

As soon as the party were fairly assembled, a horse was led into the shed—a beast with wicked eyes, and hind legs showing the unmistakable marks of contact with splinter-bars. He had been condemned as irreclaimably vicious. As he faced Mr. Adams, he stood quietly enough, watching the tiny weapon. The pistol is aimed at the forehead, right between the eyes; there is a tiny report—only one; the effect of the shock is shown to check every vital function in the frame of the animal; he sinks instantly upon his knees, and then comes lumbering down to the ground in a heap. A thin wreath of grey smoke curls from his forehead. Three or four convulsive kicks, and then complete stillness. The whole appears simple enough, and nothing more than would have happened with any bullet sent into the brain. Wait a little. The head must be examined. The grey smoke still curls from the wound as skin and muscle are removed from the skull, and then it becomes apparent that the skull is split. On handling it large pieces of bone come away easily. The surface bones are removed, and the brain beneath is found to be utterly destroyed—a mass of grey and white matter devoid of consistency. When the loose material is lifted out, there is a hole like the crater of a mine, seven inches long by six broad. Part of the bullet had been driven up to the back of the head. And this work was done by a weapon that a man can carry in his pocket!

MILITARY BREECH-LOADERS.

The principal advantages gained by applying the breech-loading principle to military arms are the attainment of rapidity of fire combined with facility in loading. A well-constructed weapon, on this principle, can be loaded easily in any position in which the soldier may be placed—either lying, running, or riding—and also gives him the advantage of the possession of three or four rifles with the inconveniences of only one. At close quarters no troops, however brave, devoted, or well-disciplined, could stand with muzzle-loaders against a corresponding force armed with breech-loaders. In the next place, we secure improved shooting. It has been urged that the rapid firing of the breech-loader will tire and unsteady men's arms; but this objection is met by the consideration that the operation of loading a breech-loader is very much less fatiguing than charging the muzzle-loader, and that the possibility of over-loading is entirely avoided.

Breech-loaders are divided into two classes, viz., those taking the consuming cartridge, and those with the metallic cartridge-case. In the former, the most noteworthy are the Chassepôt and the needle-gun; in both these the power of sustaining the force of the explosion is thrown on the breech action; the consequence of this is that an escape of gas takes place at the breech joint. In those taking the metallic cartridge-case, the force of the explosion is sustained by the base of the cartridge, which only resists this force once, but in the other instance it must resist it thousands of times.

The above two classes may be subdivided. We shall describe only those arms which have stood some Government test, or those only possessing special merit. It is surprising that no serious or deliberate attempt to introduce breech-loaders for military use should have been made until about the year 1864, when the English Government invited gun-makers and others to submit propositions for converting the Enfield rifle, and to supply 1,000 rounds of the description of ammunition which they considered most suitable to their arms. About fifty different principles were submitted to the Government, and were referred to a Select Committee.

Up to this period no Continental army had adopted breech-loaders, except Prussia and some of the minor German states, and the needle-gun was the chosen weapon. Fig. 87 represents this gun entire and divided, together with its improved cartridge.

THE PRUSSIAN NEEDLE-GUN.

This weapon appears to have been adopted by the Prussians about the year 1841, and some few were made by the British Government at Enfield, for the purpose of experiments; but, on trial, were found unsatisfactory, and breech-loading rifles for military purposes were abandoned for the time. A large quantity of Enfield rifles were, after this, made and issued to our troops.

The original needle-gun is on the bolt principle, and resembles a common door-bolt in the breech action. The first movement is to withdraw the needle by the thumb-trigger; the knob or bolt is then moved one-eighth of a circle to the left, then drawn backwards, which leaves sufficient opening to receive the cartridge; by reversing this motion the cartridge is pushed forward; the thumb-trigger is also pushed back. The piece is then ready for firing. The bolt is hollow—it contains the spiral spring and needle, from which the gun derives its name, and by which the explosion of the charge is effected. By pulling the trigger this needle is shot forward into a patch of detonating composition in the centre of the cartridge. The bullet is enclosed in a papier-mâché envelope, which answers the purpose of a wad, being about a quarter of an inch thick at the base of the bullet. The patch of detonating powder is placed in the rear of the wad. The needle must, therefore, pass through the charge of powder before it can reach the percussion composition. The whole of the cartridge is enveloped in strong paper, well lubricated with tallow. The bullet used in the improved cartridge is smaller than the bore; the paper covering fills up the grooves of the rifling, and pre-

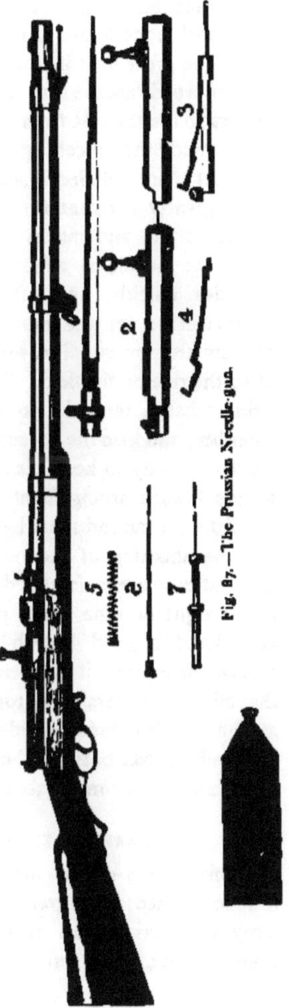

Fig. 87.—The Prussian Needle-gun.

vents windage. The grooves are four in number, and rather deep, with a moderate spiral. The size of bore is about No. 16. The great fault in this rifle is the escape of gas at the breech, which escape is so inconvenient to the soldier that he prefers to fire the rifle from the hip. Although the end of the bolt is made to fit nicely against the breech-end of the barrels the joint is not sufficiently secure to prevent the escape of gas, which greatly increases as the rifle gets foul; it also clogs the breech-action, and makes it difficult to load. If not kept clean, the needle is affected by the action of the powder, corrodes quickly, and often breaks, rendering the rifle unserviceable for the moment. To remedy this evil the soldiers are now supplied with extra needles, which they can fit to the rifle on the field. The cartridge is bad in principle, being what is termed a consuming cartridge. It fouls the chamber, and also the breech-action, and these are the parts most necessary to keep clean, to ensure the proper working of the breech arrangement. Should the cartridge happen to misfire, a ramrod must be used to extract it.

The shooting of the needle-gun is not at all accurate. The extreme range is stated at 800 yards, and for rapidity about eight or nine shots per minute. Although it is not considered a good breech-loader, it is far superior to a muzzle-loader, and it has been the means of convincing all the military Powers of Europe of the necessity for arming at once with breech-loaders. The original needle-gun weighed 15 lbs., but it has undergone some little improvement, and is now made about 11½ lbs.

SHARP'S BREECH-LOADING CARBINE.

This is one of the first American breech-loaders. It has been used with great success by the United States army in Mexico, and also during the Civil War. This breech-loader, when first introduced, was used with a made-up linen cartridge, ignited by a percussion cap, and afterwards by an improved primer, called "Maynard's." This was placed in front of the cock, and worked by the hammer in the act of cocking. This increased the rapidity of the arm—it could be fired ten times per minute. The breech-action is a dropping-block; by pressing the trigger-guard, which is fitted with a hinge-joint, downwards, one end of the breech-block is depressed, and allows the cartridge to be inserted in the chamber; the top edge of which is sharpened, so as to cut off the end of the cartridge, and expose the powder to the igniting flash which passes through the block. This breech-loader was submitted to the American Board of Ordnance at Washington, in November, 1850, who declared it to be superior to any other arm loading at the breech that had, up to that date, been submitted to them.

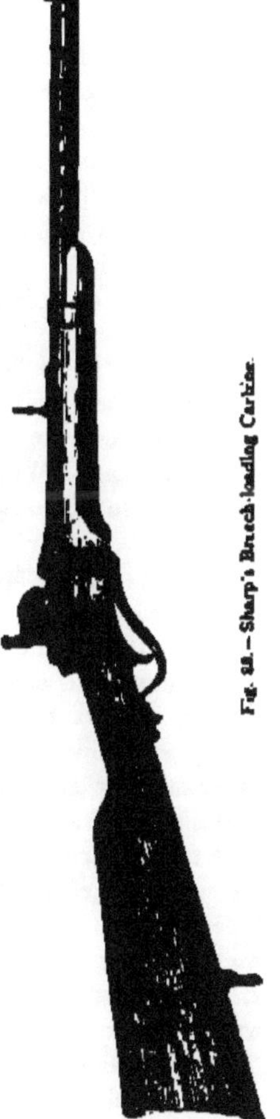

Fig. 28.—Sharp's Breech-loading Carbine.

Some of the English cavalry were also supplied with Sharp's carbine in 1857, but it was found objectionable on account of the escape of gas at the breech. To such an extent did it escape, that it would burn through a handkerchief if tied round the breech-joint. There is no arrangement to prevent this escape, which is most inconvenient to the soldier, the liability to clogging making it difficult to load after a number of shots have been fired. The Sharp's Company have lately improved this arm by making it take a metallic central-fire cartridge. The appearance of the rifle remains much the same; it is fitted with an exploding pin, and an extractor. The principle of the rifle is the same as the well-known Henry's. Fig. 88, page 167, represents this carbine, which is 500-bore, and 22 inches long in the barrel. It is also made longer and fitted with a sword bayonet for infantry use.

GREEN'S BREECH-LOADING CARBINE.

Three other kinds of breech-loaders were introduced in our cavalry for experimental purposes—"Green's," "Terry's," and "Westley-Richards'," all on the capping principle, a soft cartridge being used with a felt wad at the base.

Terry's rifle was favourably received at that time. It is constructed upon the bolt or plunger system, the cartridge being inserted at the side, and sent home by pushing the bolt forward, and secured by turning it about a quarter of a circle. It has the ordinary rifle lock and nipple. A strong percussion cap being used, the flash penetrates the paper cartridge and ignites the powder. The improvement in this rifle consists in placing a felt wad behind the powder, next the breech-bolt; the force of the explosion expands this wad, and effectually prevents the escape of gas at the breech. The bore is about the same as the Enfield. The wad remaining at the breech is pushed forward by the next cartridge being inserted; this answers as a lubricator. The wad, being greased, keeps the chamber and barrel clean. This description of breech-loader was the first which prevented the objectionable escape of gas at the breech-joint.

It had a soft cartridge, which could be made up by the soldier himself in cases of emergency.

WESTLEY-RICHARDS' CAPPING BREECH-LOADER.

This was adopted as a cavalry arm in 1861. The principle is the same as Green's and Terry's, with some difference in the breech arrangement; instead of the bolt-plunger being drawn back for loading, it is attached loosely to a hinged flap or lid. By raising this flap to a vertical position the cartridge is inserted, being pushed home by the thumb. The flap is then shut, and the point of the plunger fills up the opening in the breech; the rear end gets a solid bearing at the breech, which enables it to receive the recoil without any danger of flying open. The bore is 52. It is very accurate, and has a long range. It can be fired six or seven times in one minute with comparative ease; it is the best capping breech-loader that has ever been produced. It is much in favour as a sporting breech-loader, and will continue to be extensively used in out-of-the-way places, where metallic cartridges are difficult to be obtained. Figs. 89 and 90, page 171, represent it.

THE MONT-STORM BREECH-LOADER.

This is one of that class of breech-loaders known as chamber-loaders: a sort of muzzle-loader, which can be loaded by hand without a ramrod. The charge is deposited in a short chamber, instead of being rammed down the barrel. The breech-joint is in front of the charge, there being a hinge upon the top of the barrel. The breech is about $2\frac{1}{8}$ inches long, and turned over on to the top of the barrel to receive the charge. The block is then turned down; it is then held in position by a bolt shooting out of the break-off, which bolt is attached to the tumbler; when the cock is on the nipple it secures the breech-block. The escape of gas is prevented by an expanding ring or thimble, which effectually closes the Joint at the breech. This system has been absorbed, as it were, into the "Braendlin-Albini," and now takes the Boxer cartridge. We give a representation of the Braendlin-Albini rifle at page 179 (Fig. 94). The Mont-Storm rifle appears to have failed on account of the unsuitability of the skin cartridge. Such arms need a cartridge so thin that the fire from the cap shall pierce it, and, at the same time, the cartridge must be so entirely consumed or carried out by the discharge as to leave no residuum, to cause danger or interfere with loading. This rifle failed under the strain of the proof-charge; the hinge, and small bolt by which the chamber is locked, were broken. There are numerous breech-loaders of this class before the world, but the above description will be sufficient to explain the principle. We shall proceed to describe the more improved breech-loaders taking the metallic cartridge.

THE SNIDER BREECH-LOADER.

The English Government having decided on converting their large stock of Enfield rifles into breech-loaders, they offered a premium for the best breech-action suitable for conversion; also for the best form of cartridge adapted for this converted Enfield. In August, 1864 (as we have before stated), an advertisement was issued inviting gun-makers and others to submit propositions for conversion of the Enfield rifle, and to supply 1,000 rounds of ammunition suitable for their arm and to the service. Fifty different systems of conversion were shortly submitted, and referred to the Ordnance Select Committee. These were weeded, and eight systems only were selected for trial—viz., Storm's, Sheppard's (first system), Westley-Richards', Wilson's, Green's, Joslyn's, Snider's, and Sheppard's (second system). The first five of these were capping arms, the last three were adapted for cartridges containing their own ignition. Of the non-capping arms, one (Joslyn's) underwent no trial, the rifles converted

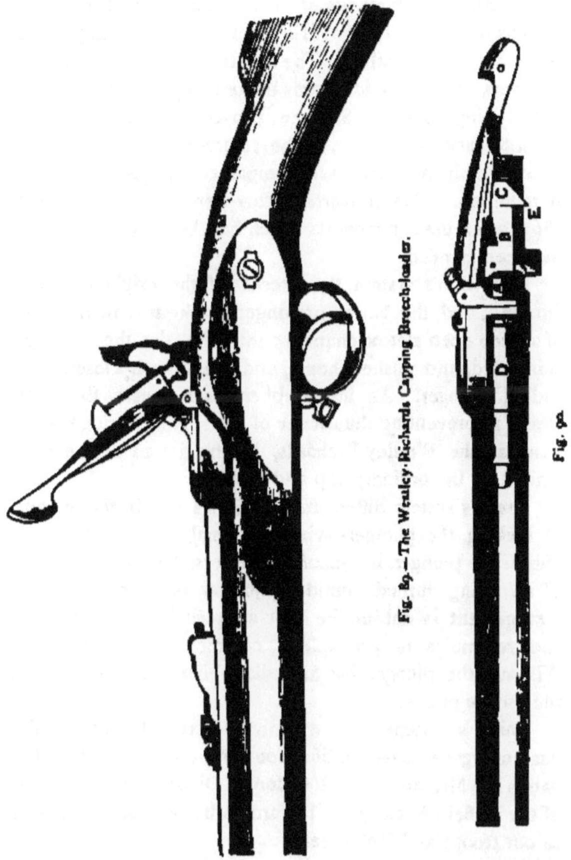

Fig. 89.—The Westley-Richards Capping Breech-loader.

Fig. 90.

on this plan not arriving in time from New York to take

part in the competition; and a second non-capping rifle, Sheppard's (second system), was rejected at the post owing to the dangerous character of the ammunition, and to imperfections in the arm itself. His capping system also proved, at the outset, unsuitable. The number was then limited to five arms, of which Snider's was the only representative of the non-capping system. Six new, proved Enfield rifles were given to each competitor, to be returned converted to his system within two months, accompanied by 1,000 rounds of ammunition. The converted arms were delivered early in 1865, when the experiments which had been determined on were commenced.

In Wilson's system the breech of the original arm is removed, and the barrel prolonged backwards in the form of a deep open slot or channel; into this slot the cartridge is inserted, and pushed home, and the breech closed by a sliding plunger. An india-rubber washer near the front assists in preventing the escape of gas. The cartridge is the same as the Westley-Richards, having a wad at the base ignited by the ordinary cap and nipple.

Green's system differs from Wilson's only in the method of locking the plunger—viz., by a small lever at the back end of the plunger, in lieu of a transverse bolt, which admits of its being turned round a quarter of a circle. This arrangement is not unlike that adopted in the needle-gun, and resembles in principle a common door-bolt. Like Wilson's, the plunger has an india-rubber washer to prevent the escape of gas.

Snider's System.—We will now describe this arm, which has undergone some modifications and improvements at the hands of Mr. Snider and Colonel Dixon, Superintendent of the Enfield Factory. This arm is being rapidly supplied to our troops and Volunteers.

About two inches of the upper part of the breech-end of

the Enfield barrel is cut away at the top, for the admission of the cartridge and bullet, which are pushed forward by the thumb into a taper chamber, formed by slightly enlarging the breech-end of the barrel. The vacant space behind the cartridge is now closed by a solid iron breech-block, which fills up this hollow and gives the rifle the appearance of a muzzle-loader. This breech-block is hinged upon the right side of the barrel, and is opened sideways by the thumb of the right hand. This block forms a false breech, and receives the recoil from the base of the cartridge. A piston or striker passes through this breech-block, the point being flush with the face of the breech, and immediately opposite the cap of the cartridge, until a blow from the hammer upon its other end, which projects above the breech, and is kept in position by a sloping nipple, drives it forward and strikes the cap, denting but not penetrating it, but with sufficient force to explode the cartridge. The empty cartridge-case is withdrawn by a claw extractor attached to the breech-block, which, when open, is drawn back with it about half an inch. The cartridge-case is brought entirely out of the barrel, and by turning the rifle sideways the empty case falls to the ground. There is a spiral spring fitted upon the hinge-rod, which takes the block and the extractor back into position. The delay caused by the withdrawal of the empty case is very slight. Some objections have been made to the spiral spring in this rifle, but it is not absolutely necessary to the system, as the first arms were made without springs; but such as have springs are much quicker and more convenient to handle. The exploding pin is also worked by a spiral spring.

In some of the first conversions a portion of the barrel was removed to admit of the breech-block; but it is now found more convenient to make the whole of the breech arrangement separate, and to screw it upon the back end of

Fig. 91.—The Snider Cavalry Carbine.

the barrel, which is shortened and screwed for the purpose. One of the advantages of this principle of conversion is, that the stock is not cut away or weakened by the introduction of the breech action. This is considered an important point, as the soldier requires a stout pike as well as an efficient fire-arm. Several of the other competing systems were rejected in consequence of the cutting away and weakening of the stocks.

The Snider proved to be the quickest and strongest breech action submitted, and was least liable to injury from the effects of of the explosion. The first arms had no arrangement to fasten down the block, except a small spring stud at the back of the breech-block, which merely kept it steady, and prevented it flying open when being handled. When a proper cartridge was used there was no tendency to cause the block to fly up, as the force is all exerted backwards; but it was afterwards found that some of the breech-blocks blew up, in consequence of a number of imperfect cartridges having been supplied. These cartridges permitted an escape of gas through the rim, this rushed under the breech-block and forced it open; this defect was remedied for the time by the issue of more perfect cartridges. The latest improvement is the spring bolt, which effectually secures the block, and prevents it from rising even when a bad cartridge is inserted.

We give two illustrations representing the "Snider."

The bore is 577, or 25 gun-gauge. This form of breech arrangement is not original. Many specimens, similar in

Fig. 92.

Fig. 93

construction, may be seen in our museums of ancient arms. The success of this breech-loading system is entirely owing to the adoption of the metallic form of cartridge. None but a perfectly gas-tight cartridge would answer with this action. The first Snider rifles submitted to competition took a paste-board cartridge with a metallic base on the central-fire or "Pottet" principle, similar to those now used in sporting breech-loaders. Although this cartridge was not exactly suited to this particular weapon, it proved to the Select Committee that a cartridge containing its own ignition, and being at the same time thoroughly gas-tight, was preferable to all other systems. Having decided to adopt the Snider, they turned their attention to the cartridge, and referred the matter to Colonel Boxer of the Royal Laboratory, who took the subject in hand, and, after great labour and many experiments, succeeded in producing a cartridge so well adapted for this rifle that it has made it a decided success.

The Snider rifle has been subjected to the following severe tests by the Government, viz., the breech has been filled with sand, water has been thrown upon it, the weapon has then been exposed to the atmosphere for a week together, taken up without cleaning and five hundred rounds have been fired from it; the same process has been repeated over again with the same results.

The expansion of the metal case tends to keep the chamber and breech action free from gas. The bullet being made of soft lead, and hollow at the base, expands freely also, and fills up all the grooves; the beeswax coating on the bullet prevents fouling or leading to any considerable extent. The number of rounds that can be fired from this rifle without cleaning would scarcely be credited, thus proving the great superiority of the system over that of the Prussian needle-gun. The thousands of times that the Snider has been fired at Woolwich without needing repair of any kind, afford a sufficient proof of the durability of the system. The range and accuracy will be noticed in another page. For rapidity the Snider was greatly superior to any other arm submitted for trial, being from eighteen to twenty times per minute. The durability of this weapon has been thoroughly established, 30,000 shots having been fired from a single rifle without effecting its efficiency.

We believe that the whole of the English army and most of the volunteers and militia regiments are now supplied with the Snider. The Turkish Government have also adopted it as the national arm. The Dutch and Portuguese are using the Snider extensively. The French have also converted a great number of their old muzzle-loading rifles on a modification of the Snider plan, which they call the Tabettier rifle. It takes the Boxer cartridge, No. 12 gauge, with a short conical bullet, which is hollow from the base to nearly the point, and filled up with a plug of papier-mâché This is done to lighten the bullet and cause it to expand so as to fill up the grooves of the rifling. The breech action differs slightly from the Snider principle, being cut away at the back of the shoe, to admit of the cartridge being inserted readily. The block takes its bearing on the two sides of the barrel or shoe. The main defect in this plan is, that if a bad cartridge should cause the block to blow open, the head of the soldier would be in danger. This could not happen in the Snider, as there is a solid standing breech behind the opening for the cartridge, which receives the force of the explosion and answers as a shield.

The Russian Government have also converted a large number of rifles on a very similar principle, without remedying its faults.

THE BRAENDLIN-ALBINI BREECH-LOADING RIFLE.

This rifle was adopted by the Belgian Government, after a series of trials which took place from the 8th to the 30th of March, 1867; the Committee decided to adopt the

Braendlin-Albini in preference to the other arms that were submitted to trial. The principal reasons which induced the Belgian Government to adopt this rifle are as follow :—

1st. This system was applicable to the conversion of their muzzle-loader.

2nd. The action was simple and solid. The construction of the different parts of which it is composed, and their putting together, being easy.

3rd. The arm is symmetrical, and does not present any inconvenient or ungraceful projection.

4th. The manipulation of the breech arrangement is quick and easy, and requires no teaching. A drilled soldier can fire twelve rounds per minute from the pouch, aiming so as to place the twelve bullets in a target fifty metres distant (the size of a man).

5th. The breech arrangement can be easily taken to pieces by the soldier.

6th. The barrel and breech can be cleaned without the necessity of being taken to pieces.

7th. During the whole course of the trials the action of the breech apparatus has not been affected by exposure of the arms to the air and rain during two nights, by repeated immersion in dirty water, by sand or dust with which the action was covered, nor by a fire of 300 consecutive rounds per arm fired in two hours and three-quarters.

8th. The closing of the breech presents all the security to be desired, and any accidental discharge is not to be feared. Five consecutive rounds were fired with very defective cartridges, in such a manner as to cause them to burst in the barrel, and they did not produce any appreciable effect on the breech arrangement.

9th. Finally, the conversion of the arms on this plan would cost less, and would be done more rapidly than on the other systems submitted.

We give a representation of this weapon at Fig. 94.

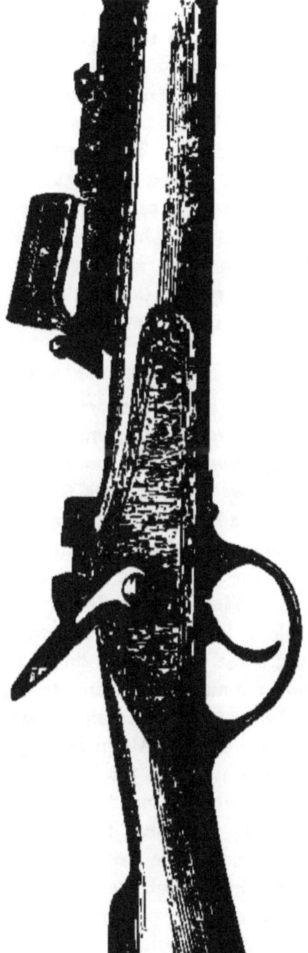

Fig. 94.—The Braendlin-Albini Breech-loading Rifle.

In appearance this rifle is like the Mont-Storm. The block is solid, and is turned over on the top of the barrel for loading. The cartridge is then inserted; the block is then replaced, being held in position by a spring stud until the hammer falls. The hammer carries with it a locking bolt, which passes through the break-off into the breech-block, preventing it from rising at the moment of discharge. This bolt receives the greater part of the recoil, which is more upwards than backwards. The hinge-joint, which is fixed upon the top of the barrels, is merely to carry the breech-block and the extractors, which are fitted on the hinge-pin, on each side of the barrel. When the breech-block is turned upon the top of the barrel, the claws of the extractor are made to eject the empty cartridge-case. The cartridge is exploded by a needle or striker, which is fixed in the breech-block, and is not visible externally when the block is down. The needle is struck by the locking-bolt, which is worked by the hammer. The Boxer cartridge, or a solid metal case, the same as the Berdan, is suitable for this weapon. The bore is small, being only 443 in the Belgian arm—that is, seven decimals smaller than the Martini-Henry rifle, but it is usually made by the patentees 577, to take the Snider cartridge.

THE FOSBERRY RIFLE.

This rifle is, in construction, similar to the Albini. The block is hinged forward, and turns over the barrel. It is not opened, however, as in the Albini rifle, by raising a hand-grip, but by drawing back the handle which is fixed to a slide on the right side of the barrel, below the breech-block. The movement of this handle and slide is parallel to the axis of the barrel, and takes effect simultaneously at two points; an incline or wedge at the end of the slide starts the block from its position, and the handle, acting on a curved lever attached to the block, completes the motion, throwing it rapidly open and setting the extractor in action at the same time. The breech-block is locked on its return, as in the Braendlin-Albini gun, by a bolt; this bolt is acted upon, however, by the tumbler itself, and not by the hammer—which is not a striker at all, but merely a means of cocking the rifle. The blow is transmitted from the locking-bolt to the cartridge by means of a piston passing through the axis of the breech-block. The ammunition is either the Boxer or Berdan brass-drawn cartridge. This system is very suitable for converting muzzle-loaders; it answers for either large or small bores. We understand that the Russian Government have adopted this plan for converting a large number of their muzzle-loading arms.

THE FRENCH CHASSEPOT RIFLE.

This arm is, in principle, the same as the Prussian needle-gun, but is certainly an improvement both in the action and barrel. The piece is cocked by the thumb, as is the needle-gun; the bolt is then turned one-quarter of a circle to the left and drawn back; the cartridge is put in and pushed home by the bolt; this bolt is turned back one-quarter of a circle to the right; the piece is then ready for firing. It is loaded when at full cock. To put the rifle at half-cock, the bolt must be turned only about one-eighth of a circle. To do this the trigger must be gently pulled, which will allow the cocking arrangement to enter a slot in the bolt-plunger. This movement effectually locks the breech-action, and answers the purpose of a safety bolt. It can only be fired by bringing it to full cock with the thumb and turning the bolt completely down. The point of the bolt enters the barrel, and is fitted with an india-rubber washer, which partially prevents the escape of gas. Fig. 95, page 182, represents this rifle, with the breech-action in separate parts; also the cartridge, which is differently constructed to that of the needle-gun, the ignition being placed in a percussion cap near the base. The needle enters the cap, which is so placed that the opening is towards the breech arrangement, thus the needle strikes the inside of the cap instead of the outside. The great improvement in the Chassepot is the barrel, which is a small-bore (434), made of steel, and takes the large charge of 85 grains, and a solid conical bullet (weighing 380 grains). The extreme range is 1,800 yards; the weight of the rifle complete with the sword bayonet is 11 lbs.

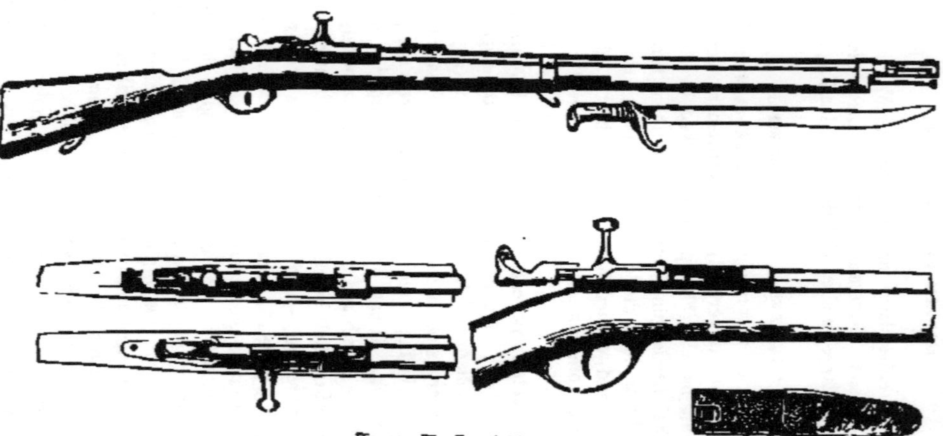

Fig. 95.—The French Chassepot.

The barrel is rifled with four deep grooves, having the spiral to the left instead of (as is usual) to the right, with one turn in 21¾ inches. The reason for rifling with the spiral to the left is to counteract the pull of the trigger, which is a very bad one—it drags dreadfully, and requires a long pull to discharge the rifle.

The bullet for the Chassepot is made larger than the bore of the barrel, is driven through it, fills up the grooves and prevents windage. But there are several serious objections to this plan. In the first place, it occasions great friction, and much recoil; the barrel also leads very quickly; the bullet leaves the barrel nearly square, which is a bad form for any projectile. For perfect accuracy of flight the bullet should be as nearly cylindrical as possible, and the more even the surface is the better. The Chassepot bullet is quite opposite to this, and resembles the Jacop shell after leaving the barrel.

The cartridge being what is called self-consuming, there is a great escape of gas at the breech; this causes such an accumulation, that after the firing of a number of shots, the manipulation of the breech is impossible without its first being cleaned. There is no lubrication in the cartridge, It is said that the Frenchmen spit on their cartridges, force their fingers into the breech action, and give every possible sign that, after a few shots, the Chassepot gets so foul that they do not know how to treat it. There is a difficulty in getting the cartridge into the chamber when the rifle is foul; if force is used it becomes dangerous—the cartridge being soft, the percussion cap is compressed between the bullet and the point of the bolt, and has been known to explode in the act of loading, in many cases injuring the hand of the soldier severely. However, with all these imperfections, the Chassepot has proved a most deadly weapon, and it appears to have served the French well in all their engagements with the Germans during the recent war, especially at the longer ranges. Although we consider this a very imperfect military weapon, it has proved, during the late campaign, that an imperfect breech-loader is infinitely superior to the muzzle-loader.

THE BURTON RIFLE (NO. 2).

This rifle is constructed on the bolt system. The arrangement of closing the breech is similar to that of the Prussian needle-gun. The locking of the bolt is effected by means of a small projecting boss, on its upper side. On the bolt being pushed forward by a lever handle, it passes through a slot in the back part of the shoe, and is then turned to the

Fig. 96.—The Carter-Edwards Breech-loading Rifle.

right, preventing the plunger from being withdrawn until the boss is once more brought opposite to the slot. The difference between the Burton rifle and the needle-gun is, that the former is adapted for the metallic cartridge-case.

Mr. Burton has another rifle, which is constructed according to the breech-block system.

THE CARTER-EDWARDS BREECH-LOADING RIFLE.

This rifle is an improvement upon the Chassepot. The breech arrangement is on the same principle, but adapted for the metallic central-fire cartridge.

The empty cartridge-case is extracted ("ejected"), and the rifle loaded and cocked, in two motions: turning the handle one-quarter of a circle, drawing back the bolt, cocks the piece and ejects the cartridge-case at the same time; a new cartridge is then inserted, the bolt is returned, and the rifle is ready for firing.

The action of the lock is a piston instead of a needle, which is undoubtedly an improvement, inasmuch as it is not subject to the corrosive action of the gases generated by the explosion. The metallic cartridge-case entirely prevents any escape of gas at the breech. We give an illustration representing this arm at page 184, Fig. 96.

THE WILSON BREECH-LOADING RIFLE.

This rifle is also on the bolt principle, and adapted for the metallic cartridge. At the end of the bolt is a knob, which is gripped by the thumb and forefinger. By turning it one-quarter round the bolt can be drawn back for loading, and this same motion extracts the empty case; the bolt being returned, the rifle is ready for firing.

THE KERR BREECH-LOADER.

This may be classed as one of the best bolt actions. The breech arrangement is similar to the Burton, but retains the ordinary lock and hammer, and takes the Boxer cartridge.

THE OBJECTIONS OF THE BOLT SYSTEM.

The Special Committee on Small Arms rejected all guns constructed on the bolt system, as they considered that the process of thrusting forward the bolt against the rear of a loaded cartridge was a source of danger. We could give descriptions of a large number of guns constructed on this principle, but think those we furnished will be sufficient for all practical purposes.

The bolt system was thoroughly tested by the Government Select Committee, and during one of these trials an accident happened to a member, causing him serious injury.

Fig. 97.—The Spencer Carbine.

We have ourselves witnessed a similar accident with a gun on the bolt principle, caused by a bad cartridge, having split at the base, allowing an escape of gas into the breech action, which blew the bolt-plunger completely over the head of the shooter, severely injured his hand, and split the stock of the rifle. We consider the bolt principle in every way bad. In loading rapidly the base of the cartridge receives a sharp blow, which is often sufficient to explode a cartridge containing a sensitive cap. The concussion would do this without the cap having been struck.

THE AMERICAN REPEATING OR MAGAZINE RIFLES.

The Spencer appears to have been the first successful rifle constructed according to this plan; it was patented on the 6th of March, 1860, in the United States. We give a representation of this rifle at page 186, Fig. 97. The magazine is in the butt. To load, the muzzle is pointed downwards, the magazine lock is turned to the right, the inner magazine tube is withdrawn, the cartridges are dropped into the outer magazine, ball foremost, then the tube is inserted and locked. There is a spiral spring fitted in the magazine, which forces the cartridges up to the breech chamber. The first cartridge is forced forward into the chamber of the barrel by moving the guard-lever downwards, as shown by the engraving, and immediately drawing it back. It can be loaded with the hammer down, but should be kept at half-cock while the cartridge remains in the chamber. To fire, bring the hammer to full cock, and by pulling the trigger, it strikes the percussion slide, forcing it against the rim of the cartridge, and exploding it. The discharged shell is withdrawn by the opening motion; there is a carrier block that moves the shell-drawer over the cartridge guide, which is then depressed by a spring. This same guide aids in conducting the new cartridge to the chamber. We give a drawing of the cartridge (which is rim-fire) for this rifle, full size. The gauge is 500. The charge of powder is small; the weight of the bullet is 1 oz. for the military, and ⅞ oz. for the sporting rifle. It can be fired seven times in ten seconds, but only fifteen times in one minute; it can also be used as an ordinary breech-loader. This rifle was used with great success in America, during the Civil War. It was considered that one man armed with the Spencer, was equal to five or six armed with muzzle-loaders.

THE HENRY REPEATING RIFLE.

This rifle has a magazine under the whole length of the barrel, and contains fifteen charges. The gun is manipulated in two motions; it can be loaded and fired thirty times a minute. By depressing the lever-guard, and bringing it back quickly, the old case is extracted, the rifle is cocked, and a new cartridge inserted in the chamber. The magazine is a rather delicate arrangement; it was improved upon considerably and superseded by the Winchester rifle, of which we give a description.

THE WINCHESTER REPEATING RIFLE.

This rifle, so far as the mechanism for loading and firing is concerned, is precisely the same as the Henry, except in the form of the cartridge extractor and magazine. The latest improvements consist in an entire change in the magazine, and the arrangements for filling it. By these changes the gun is made stronger and lighter, the magazine is closed and strongly protected, is more simple in operation, and requires fewer motions to fill.

The magazine is a tube containing the cartridges, which are placed under the barrel, protected by the wood forend of the stock. The cartridges are put in base first, and forced up to the breech by means of a spring. It is impossible to put the cartridge in the wrong end first. The magazine can also be replenished at the breech-end, and that without changing the normal condition of the gun. One advantage is that it can be loaded as an ordinary breech-loader, and fired thirty times per minute by an expert.

The principal novelty in this gun is the magazine, and the manner of loading from it. It consists of a tube under the barrel, extending its entire length, of sufficient diameter to admit the cartridge freely. A section of this tube, near the muzzle, contains a spiral spring to throw the cartridges upon a carrier block in the rear.

When the spring is pressed into this section, it turns upon the axis of the bore, leaving the magazine open for the introduction of cartridges, of which it holds fifteen.

On closing it after firing, the spring throws a cartridge

Fig. 98.—The Winchester Repeating Rifle.

upon the carrier block, which by a forward movement of the trigger-guard is raised to a level with the chamber; the hammer by the same movement being carried to full cock. A reverse movement of the guard bringing it to its place again, forces the cartridge into the chamber, and the gun is ready to fire.

The ammunition is fixed, metal-cased, with fulminate or cap, in the rear; the hammer upon falling strikes a rod or breech-pin, on the front of which are two sharp points, which are driven into the rear of the cartridge, thus exploding it. The weight of the rifle is about 9½ lbs.; the bore is 42-100ths of an inch. We give an illustration showing the internal mechanism at page 189, Fig. 98.

An arm like the Spencer and Winchester can be used as a repeater or simple breech-loader at will, by turning off the magazine and inserting a cartridge in the usual manner. When hard pressed the magazine can be drawn upon, and fifteen shots can be delivered in fifteen seconds.

The direction in which repeaters err is complexity of construction. If this difficulty could be overcome, the advantages over the breech-loader would be considerable, not merely for the cavalry and artillery, where an intensely rapid fire is generally required for a few decisive moments, but for the universal equipment of troops. But it is not probable that repeating arms will become general in their present complicated form; their liability to get out of order more than counterbalances the advantage gained by rapidity. Durability is one of the first considerations in a military breech-loader. Long range and low trajectory, combined with accuracy, are other important qualities which the repeating arms do not possess, owing to the small charges of powder and short bullets that are used in the cartridges, which conditions are necessary to admit of the quantity to be carried in the magazine. The breech action of these rifles is not calculated to stand the strain of heavy charges, such as are used in the Martini.

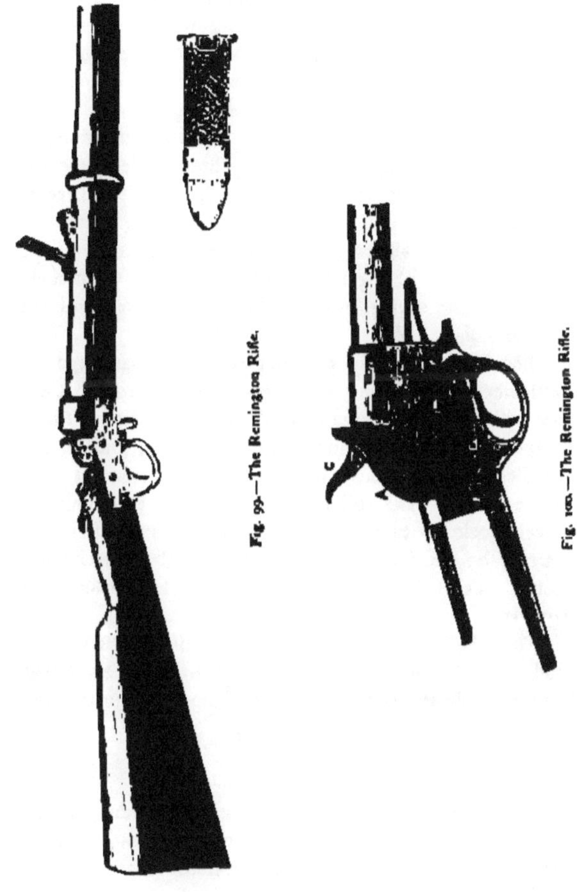

Fig. 99.—The Remington Rifle.

Fig. 100.—The Remington Rifle.

THE REMINGTON RIFLE.

This rifle was tried at Wimbledon, as long ago as 1866, and attracted considerable attention at that time, in consequence of the extraordinary rapidity with which it was loaded and fired—as many as fifty-one shots were discharged from it in three minutes, but the shooting was of course very wild. We give two illustrations representing this rifle at page 191, and by reference it will be seen that the action consists mainly of two pieces, one being the breech piece and extractor, and the other the hammer breech-bolt. This breech-piece and hammer-bolt each work upon a strong centre.

The letter A shows the breech-piece closed; B C, the hammer down with the breech-bolt backing up the breech-piece; D, the spring holding the breech-piece until the hammer falls.

The bore is usually 500, but it is also made 450 for bottle-necked cartridge. At the trial above referred to it was shot with the small charge of 1½ drams of powder. It is at present made to take a much larger charge (85 grains), and the Berdan cartridge. The breech arrangement is very simple, but it lacks solidity. The method of holding the breech-block up to the barrel is quite original, but it can scarcely be considered truly scientific, as the breech-piece should receive its support from immediately behind the cartridge.

This rifle is made in America by machinery, and can be produced in large quantities at a moderate price. It has been extensively used in America, France. Denmark, and Austria, and also by the papal troops.

THE BERDAN RIFLE.

This rifle is the invention of General Berdan, of the American army; it is a combination of the Braendlin-Albini and the Chassepot. We give an illustration representing it at Fig. 101.

There is a hinged block which turns over the barrel and extracts the cartridge case. It is locked in position for firing by a bolt resembling the cock of the Chassepot. The lock is worked by a spiral spring. The blow given by the locking bolt is communicated to a striker working in the breech block. It is the Braendlin-Albini, with the lock in the centre instead of on the side. This gun is used by the Russian Government. It is open to an objection. When loading rapidly the needle is liable to get against the cap and cause a premature explosion, there being no spiral spring to force back the needle. It is a great fault in all breech-loaders constructed on this principle — they ex-

Fig. 101.—The Berdan Russian Breech-loader.

pose the base of the cartridge to a dangerous amount of concussion in the act of closing the breech. The bore of the Berdan rifle is small: it takes the Berdan metallic cartridge case.

THE ROBERTS BREECH-LOADER.

This rifle was adopted by the United States Government for the conversion of the Springfield rifle. It is a simple arrangement. The parts are only five in number. The first piece is an iron breech frame or shoe screwed upon the barrel. This shoe is embedded in the stock in the place of the old breech pin; the breech block is inserted in this shoe, and supported against the rear end on a semi-circular shoulder forming the back of the shoe; the rear of the breech block is turned to fit with exactness this semi-circle, and is plugged round it as a fulcrum. When the breech block is in place in the shoe it forms a curved lever, the handle projecting backwards. When this handle is raised, as shown in the illustration (p. 195), the front part of the breech block is sufficiently depressed to admit of a short cartridge being inserted into the barrel. The cartridge is 1¼ inch long only. The extractor is a curved lever fixed on the left side of the chamber, with one arm behind the flange of the cartridge case, and the other operating in a vertical groove on the left side of the breech block. The cartridge extractor is acted upon by the breech block when it descends below the chamber of the barrel.

The firing pin is situated on the right side of the breech block, and is struck by the ordinary musket lock. It is made to take either the central-fire or the rim-fire cartridge. The principle appears to be thoroughly sound and durable.

THE PEABODY BREECH-LOADING RIFLE.

This rifle is the invention of Mr. H. O. Peabody, of Boston, Massachusetts, United States of America. The

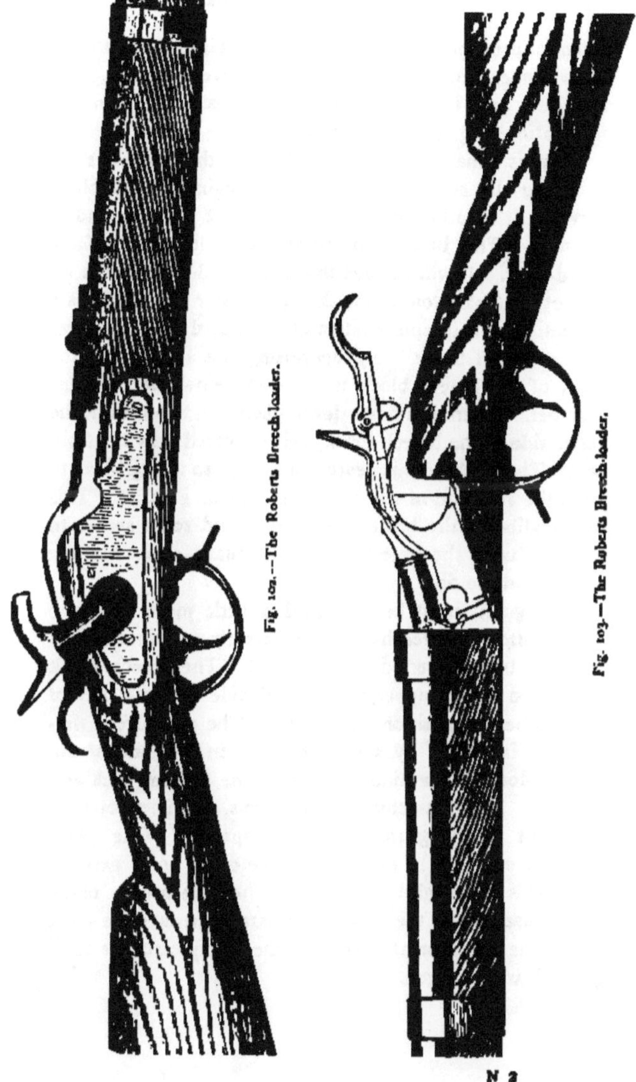

Fig. 102.—The Roberts Breech-loader.

Fig. 103.—The Roberts Breech-loader.

essential principles of the gun were conceived by him some years since, and upon these, patents have been issued in America, and in several other countries. The annexed illustration (Fig. 104) represents it.

There is a metal breech frame or shoe that connects the barrel to the stock, in which the swinging breech block moves. It is attached to and pivoted at the rear end of the shoe. This block is depressed at the front part of the guard-lever. It will be seen that this guard-lever works on a pivot at the bottom of the shoe (see letter J). There is a projecting arm in continuation of the guard which engages in the breech block. By depressing this lever-guard the point of the breech block is dropped below the cartridge chamber. There is a brace lever placed in a recess in the under side of the breech block, being pivoted near the front part. The back of this lever is a spring, so arranged as to press the rear end of the lever firmly upon a roller. This combination of the brace lever, spring, and roller serves to securely fasten the breech block and guard-lever when the arm is ready to fire.

The guard must be depressed a little more than one inch, to drop the breech block sufficiently to admit of the cartridge being inserted in the barrel. The breech block is grooved on the upper side, to coincide with the bore of barrel when in position for loading. The cartridge is rim-fire, and is ignited by a slide working in the side of the breech block. This slide is struck by the ordinary lock and hammer. The extractor has two arms, and is pivoted at the front of the guard, projecting upwards. One point takes the cartridge by the rim, and the other point extends backwards and slightly upwards. When the breech block is depressed below the bore of the barrel it strikes the short arm of the extractor, and jerks out the empty cartridge shell quite clear of the gun.

This gun was submitted as early as 1862 to the officer commanding the United States arsenal at Watertown, and was reported upon favourably.

Fig. 105.—The Peabody Breech-loading Rifle.

The Peabody rifle appears to be the first of the class having the dropping block pivoted at the rear end and above the axis of the bore. The chief advantage of this principle is, that the point of the block describes part of a circle, and moves clear of the base of the cartridge at the moment it is depressed, thus effectually preventing any jamming of the breech block by the expansion of the cartridge at the base, which has been known to occur in rifles constructed like Sharp's, where the whole of the breech block slides down below the bore. There is also less friction wear and tear in the rear pivoted system.

The recoil is received chiefly upon the breech frame or shoe, immediately behind the pivot, and partly upon the arm of the lever, which has to keep up the breech block in position for firing. The breech action of this rifle has been tested by the United States Board of Ordnance, with a charge of eighty grains of powder and five balls, with no injury to the gun. The United States Government having offered every facility, and even inducement, to the manufacturers of breech-loading fire-arms, hundreds of mechanics at work in the arms-manufacturing establishments of the United States have competed with patient diligence for the honour of producing the most effective and simple weapon which, in the hands of the most clumsy and least intelligent soldier, could be manipulated, without danger to the user, and be capable of the deadliest effect upon the enemy. The result has been, that a large number of small arms of every variety have been produced, many of them presenting claims to merit, and the Peabody we class as one of the best of the American inventions.

Both the Martini and the Westley-Richards breech-loading rifles are certainly only improvements on this principle. But there is a point in which the American rifles signally fail, and that is accuracy of shooting. The trajectory of all their rifles is very high. The charge of powder used in them is too small to give a flat trajectory and long range. The cannelure bullet, which is generally used, is not at all calculated to give extreme accuracy. The English-made rifles are far superior in that respect, as the Wimbledon trials have fully demonstrated.

REPORT OF THE SELECT COMMITTEE ON BREECH-LOADING SMALL ARMS, 1868.

In answer to the Government advertisement, a large number of rifles were entered and submitted, and, after the preliminary trial, were all rejected but ten, which were retained for further trial. These consisted of the Bacon, Berdan, Carter and Edwards, Henry Kerr, Martini, Money-Walker, Westley-Richards (two systems), and Wilson. Of these guns, the Bacon, Carter and Edwards, Kerr, and Wilson represented the bolt system; the remainder represented the block. During the subsequent trials, two accidents occurred with bolt guns, one with the Bacon and the other with the Wilson; and when the defective cartridge—which had been purposely supplied for experiment with this class of gun—came to be used, the Bacon, Wilson, and Kerr rifles showed that, under these exceptional circumstances, they were capable of exploding the cartridges prematurely. The Bacon did actually explode a cartridge; the Kerr and Wilson indented the caps. This left only the Carter and Edwards; but evidence before the Committee led to their rejection of this gun. The fall of a cartridge upon the floor has been known to explode it when fitted with over-sensitive caps, or non-safety anvils. Such cartridges, jambing in the breech of a bolt gun, would, it is reasonable to assume, in a certain proportion of cases be almost certainly fired. The Committee eventually (and, we consider, very properly) rejected this class of breech-loaders altogether.

In the rapidity trials they were placed in the following order:—

		Bore.		Min.
1st.	Westley-Richards (elevating block)	·45 inch ...	20 rounds in	1.0
2nd.	Martini	·433 " ...	"	1.2
3rd.	Henry	·45 " ...	"	1.7
	Westley-Richards (falling block)	·45 " ...	"	1.7
4th.	Berdan	·45 " ...	"	1.10
5th.	Money-Walker	·5 " ...	"	1.14

In the damaged-cartridge and sand tests, all the arms acquitted themselves satisfactorily. In the exposure tests the Berdan, Westley-Richards (elevating block), and the Money-Walker became decidedly unserviceable; and the other Westley-Richards went very near to breaking down. This trial practically reduced the competition to the Henry and Martini, and of these the Martini gave the better performance, having been found in perfect order at the close of the experiment, while two springs of the Henry were broken. In all the trials after the close of the prize competition, the Martini had been fired with the Boxer cartridge case and a compressed powder charge; and to the former its very different performance, as compared with that in the prize trials, when copper cartridges were used, may be attributed. It had been by this time discovered that loose powder was preferable, as regards accuracy, to compressed powder, for ·45 bores. It became necessary to test the Martini with a longer loose-powder cartridge case, and fitted to a ·45 barrel. An arm adapted to these conditions was supplied, and passed successfully through the rapidity. sand, damaged-cartridge, and exposure tests. The rapidity attained exceeded that of any other arms, being twenty rounds in fifty-three seconds. After a week's exposure to rain and artificial applications of water, and to firing 400 rounds, at intervals during this period, the arm, uncleaned, fired twenty rounds in one minute three seconds, so that it was evident that the Martini action was equally adapted to the long or short cartridge, and it was therefore placed in direct competition with the Henry.

In the above trials—under the heads of safety and strength—both arms were considered equal; in regard to the number and simplicity of parts, the Martini has the advantage. It has only thirty parts, against forty-nine in the Henry. The extractor plate, soldered on to the barrel of

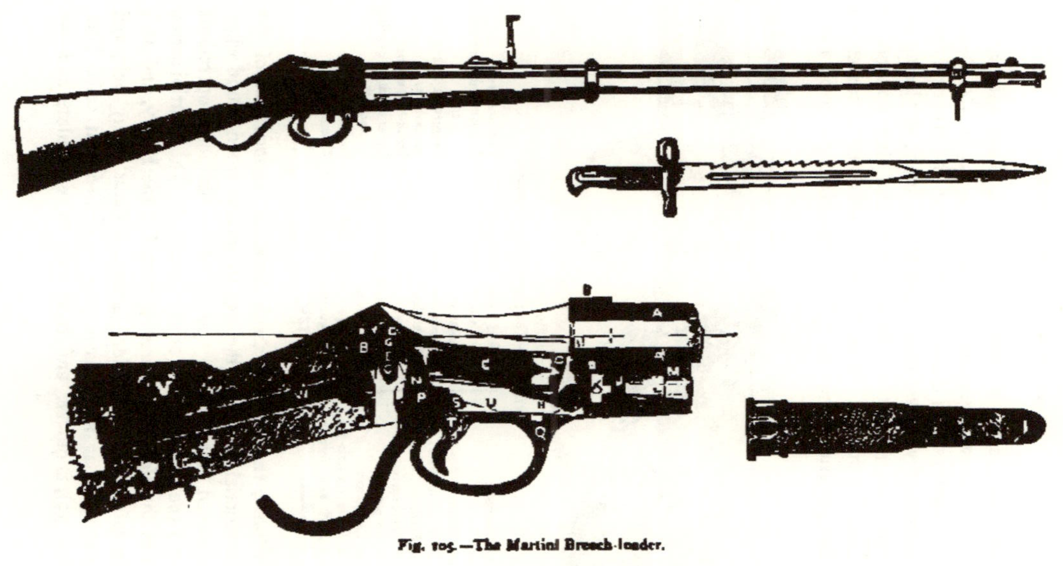

Fig. 105.—The Martini Breech-loader.

the latter arm, is also considered a disadvantage. In facility of manipulation the Martini, owing to the absence of a hammer, has the advantage; and in the Henry there is a possibility, as was discovered at the trials, of placing the cartridge in front of the extractor, and thus temporarily disabling the arm. The Martini is stated to be rather the cheaper arm of the two. The Committee preferred a gun

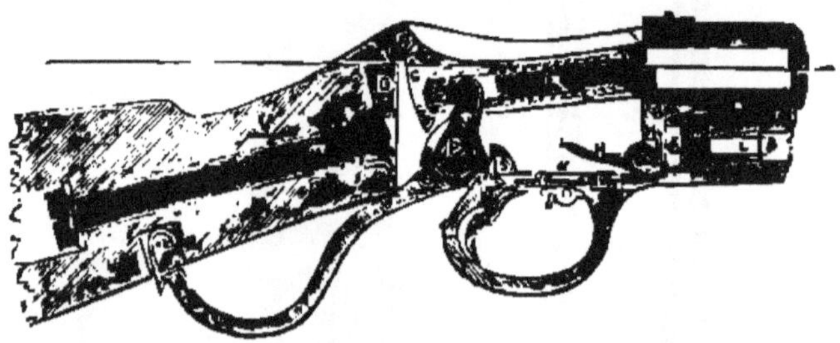

Fig. 105.—The Martini Breech-loader.

AA Barrel.
BB Body.
CC Block.
D Block axis-pin.
E Striker.
F Main-spring.
G Stop-nut.
H Extractor.
I Extractor axis-pin.
Rod and fore-end holder.
K Rod and fore-end holder screw.
L Ramrod.
M Stock, fore-end.
N Tumbler.
O Lever.
P Lever and tumbler axis-pin.
Q Trigger-plate and guard.
R Trigger.
S Tumbler-rest.
T Trigger and rest axis-pin.
U Trigger & rest-spring.
V Stock-bolt.
W Stock-bolt washer.
X Lever catch bolt, spring and pin.
a Locking-bolt.
b Thumb-piece.

without, to one with a side lock, in consequence of the occasional liability of the lock to become wood-bound when exposed to wet, to say nothing of the additional operations and the multiplication of parts entailed. The Martini action was therefore, in the end, preferred to the Henry, and as it is safer than the Snider action, without safety bolt, and stronger, has fewer parts (the Snider, without safety bolt, has 39), is quicker and more easy to manipulate, and—so they considered—costs less (which we very much doubt), the Martini system of breech mechanism was ultimately recommended for adoption for the future arm.

We give two illustrations representing the arm (Fig. 105 at page 201, and Fig. 106 at page 202).

SECTIONS OF MARTINI BREECH-ACTION.

It now only remained to wed the Martini action to the Henry barrel. The ceremony was performed at Enfield about the beginning of the present year, and four Henry-Martini arms, with a supply of ammunition, were furnished for further experiment. It may be interesting here to give the results of the final trials of the complete arm :—

Range.	Mean Figure of Five Targets of 20 Shots each.	Best Figure obtained.	Angle of Elevation.
	Feet.	Feet.	° ′ ″
300 yards.	·57	·47	0 38 34
500 „	·95	·79	1 1 26
800 „	1·63	1·29	2 2 29
1,000 „	2·80	2·19	2 38 26
1,200 „	3·46	2·28	3 55 34

Rate of fire attained, without taking aim, 20 rounds in 48 seconds.

Riflemen will know how to appreciate these figures, which represent the capabilities of the proposed weapon, and which, we hope, will be at least approximately reached when the arms are supplied in large numbers.

A very few words, added to the accompanying drawings of the new rifle, will suffice to describe the breech action.

The breech is closed by a block which swings on a pin passing through the upper rear end of the shoe, like the Peabody, the recoil being taken by the shoe. (This point has been fully established practically and theoretically—there is no strain on either the block pin or lever pin.)

The cartridge is exploded by a direct-acting piston, which is driven by the action of a strong spiral spring within the breech block. The block is acted on by a lever to the rear of the trigger-guard. The act of pushing the lever forward causes the block to fall, the spring to be compressed, and the empty cartridge case to be ejected. On drawing back the lever, the block is raised so as to close the breech, and the arm is ready for firing.

The indicator at the side shows if the arm is cocked or not; there is also a safety bolt, which secures the trigger and answers for half-cock. It was determined to give the Martini-Henry rifle a rough practical trial in the hands of soldiers before finally adopting it, and 200 arms (hand-made) were supplied to the troops. It may be interesting to some of our readers if we give in full the history of these 200 experimental rifles.

THE MARTINI-HENRY ARMS AND AMMUNITION ISSUED FOR TRIAL, 1869.

A Blue Book composed of abstract of reports on the 200 experimental Martini-Henry rifles, together with the remarks of a Special Committee appointed to consider the same, has been published.

This Committee was composed of four military men and two civilians, as follows:—Colonel Fletcher, president; Captain Haig, R.A., Mr. E. Ross; these three acted on the original Selecting Committee. The three new names are Lord Elcho, Captain Chapman, Inspector of Musketry, and Captain Aylmer. In 1869, these 200 rifles, with a quantity of the proposed Boxer-Henry ammunition, were distributed for trial at home and abroad, and "the forms of report which accompanied the rifles consisted of a series of twenty-one questions, the replies to which have been arranged *seriatim*." Abstracts of these reports, with remarks of

Committee thereon, go to make up this Blue-book, in noticing the contents of which we will take those questions first which relate more directly to the ammunition, viz., 2, 3, 17, 18, 19, and 20, noting the answers we think most deserving of attention.

Question 2.—What difficulties, if any, have been found in loading?

Twenty-three answers say none, out of forty-seven in all. Among the remainder of answers the following are recorded:—Rep. No. 8.—"If not well pressed home with the thumb, the cartridge does not always explode, though the breech is closed." No. 17 Rep., Hythe School of Musketry, says—"With No. 26 rifle, the paper round the cartridge ruffled up when putting it into the breech. The paper of cartridge also stripped with No. 137 rifle, and became jammed, preventing the cartridge from entering or being removed from the barrel." Rep. 21, Gravesend, Royal Engineers, states—"Several cartridges would not enter. This was apparently caused, in a few instances, from the base cups being too large; the majority, by the paper getting scraped up and jamming the cartridge partly into the chamber, so that the ramrod had to be used." No. 25 Rep., Portsmouth, 2nd Battalion 2nd Royal Engineers, states—"Several (difficulties), consequent on malformation of cartridges, and on the paper rucking up against the edge of the chamber." No. 28 Rep., Portsmouth, 101st Regiment, says—"In ten instances the cartridges would not enter the chamber of any of the rifles." Various other reports state that difficulties in loading had arisen from bent and faulty cartridges.

Question 3.—What difficulties, if any, have been found in extracting cartridge cases?

To this inquiry twenty-eight answers are, None. Among the difficulties reported the following appear:—No. 4 Rep.

—"In rifle No. 96 the extractor would not work; thinks it was owing to rust. Rifle No. 157 did not extract blank cartridge easily." No. 6 Rep. says—"None, except on one occasion, when the base of cartridge was extracted, the case remaining in the barrel." No. 14 Rep. is, "When the cartridges missed fire, the ramrod had to be used to get them out, the extractor not getting sufficient bite of the base of the cartridge to move the weight of the unfired cartridge." Rep. No. 29 says—"Very frequently the ejector has failed to extract the case, only moving it slightly." Rep. No. 46, Quebec, 69th Regiment, answers—"None, except in rifle No. 100. In most instances the cases had to be pulled out by the fingers."

Question 17.—Has the ammunition been found liable to become broken or damaged in transport or when carried in men's pouches?

Rep. No. 7 answers—"Some have become bent, but were straightened and then used." Rep. No. 18 says—"Some of the cartridges have become bent and dented, owing to several of the packages having been badly packed." Rep. No. 30 replies—"One or two have become bent, probably in the men's pouches. The length renders them liable to injury." No. 34 Rep., Montreal, 1st Battalion Rifle Brigade, reveals an omission. It says—"No pouches suitable for the ammunition have been issued from the Store Department."

Question 18.—Has any difficulty arisen from this cause (viz., damaged cartridges) in loading or extracting?

Twenty-seven Noes are recorded; but among the other twenty there are some damaging reports. Take Rep. No. 16—"If dented or bent, it is impossible to load." Rep. No. 23 says—"The paper is the weak part of the cartridge; it becomes rucked up and causes the case to jam. The blank cartridge is difficult to load." Rep. No. 25 states that "several cartridges damaged in transport would not enter the chamber; others required force to push them in. No difficulty in extracting." Rep. No. 27 answers—"Yes, from bent cartridges; 38 cartridges out of 2,750 would not enter the barrel, and some others were hard to force in." Others complain of "defective shape" in five or nine cartridges.

Question 19.—Is the form of the cartridge found convenient to use?

Fourteen answers all agree in their replies that it is "too long." Rep. No. 5 adds—"liable to become damaged, and entails cumbersome pouches." Others remark that it would be "more convenient," "preferable," or "better," if shorter.

Question 20.—Have any of the cartridges cut round the base?

To this inquiry thirty-three replies are—No. Other reports show only seven rounds of blank and four of ball cartridges to have been "cut" by the act of firing. Four reports prove something else—another defect not asked about. These are as follow:—Rep. No. 11.—"No actual cutting, but slight escape of gas close to base cup, and where the cap overlaps." Rep. No. 12 is similar. Rep. No. 14 answers—"No (as to the cutting, but adds), there is a slight escape of gas round the base of *most* of the ball cartridges." No. 15 Rep. is similar.

In looking at these various reports in the gross, and the faults proved to exist in the "recommended" Boxer-Henry cartridges, there is more than enough to condemn any cartridge issued for military service. We are not surprised, therefore, to learn from the remarks of Committee on the reports given in reply to Question 20, that the "long thin *spurtle*" of a cartridge that was issued for trial is summarily condemned, and a shorter one "provisionally adopted," which is also to get rid of the difficulty of the paper covering "rucking up." We will give the Committee's own remarks:—"With regard to the objections which have reference to the bending of the cartridges, and to the rucking up of the paper, the Committee would observe that a remedy has already been provided by the *provisional adoption of the short chamber cartridge*, which is stronger, and has no external covering of paper." Thus it appears that this Special Committee is compelled to condemn what the original Special Ordnance Sub-committee recommended for adoption. We may state that the condemnation of the long thin cartridges has led, necessarily, to an alteration of the breech mechanism of the rifle itself, for it comes out incidentally in the remarks of the Committee on the reports given as replies to Question 14, that there are "twelve short-action rifles now being made for issue to the troops."

There is one singular omission in the series of twenty-one questions, namely, that whilst Question 11 refers to whether the weight of the *rifle* be a practical objection on service, no question was put as to the weight of the *ammunition!*

We may say that the weight of sixty rounds of the small-bore Boxer-Henry cartridges issued for trial was about nine or ten ounces *heavier* than the sixty rounds of Enfield muzzle-loading cartridges, the first being about 6 lb. 4 oz., the second 5 lb. 11 oz.—a heavier cartridge recommended for a rifle whose rapidity of fire is about three or four to one of the muzzle-loader! As our readers may be interested in knowing the *lengths* of the several cartridges referred to, we give them:—The Boxer-Snider Enfield cartridge is 2½ in. long; the Boxer-Henry small-bore cartridge, 3¾ in. long—just 1¼ in. longer; and the length of the "provisionally adopted" crimped Boxer cartridge is the same as that for the muzzle-loading Enfield, viz., 3 in. full. This crimped Boxer cartridge looks like a rough imitation of the Daw bottle-necked cartridge, the rear-end being made the same diameter as that for the Snider-Enfield rifle, thereby gaining a reduction in length of three-quarters of an inch.

Many objections have been made to the "Martini action," chiefly on account of the spiral spring. We do not consider this to be an objection, as a spiral spring can be made to stand as well as an ordinary main-spring.

In the manipulation of the Martini, care is required. If the lever is not brought down with a sharp jerk, it will not properly extract the cartridge case, there being but a slight play in the extractor, with but very little leverage. This jerk is necessary to extract the empty case. If the lever is pulled down gently, the case is only partially extracted, and would stop the action.

The breech-block in the Martini is an admirable arrangement, and will stand a great amount of hard work; but we think the mechanism of the lock will not prove so durable, and it has already been improved on by Westley Richards.

THE WESTLEY-RICHARDS MILITARY BREECH-LOADING RIFLE.

This rifle is similar to the Martini; but the lever is pivoted in front of the trigger-guard, instead of behind. The arrangement for exploding the cartridge is different: there is no spiral spring, the ordinary main-spring is used. The hammer, striker, and tumbler are in one piece, and resemble the cock of a central-fire revolver. The nose of the cock is pointed, and strikes the cap through a hole in the breech-block.

The block is supported at the extreme end, next the cartridge, by an arm of the lever, which is in a vertical position. The extractor is a powerful lever, capable of extracting a tight cartridge-case and throwing it clear of the rifle.

The safety bolt is a cam on the side of the breech

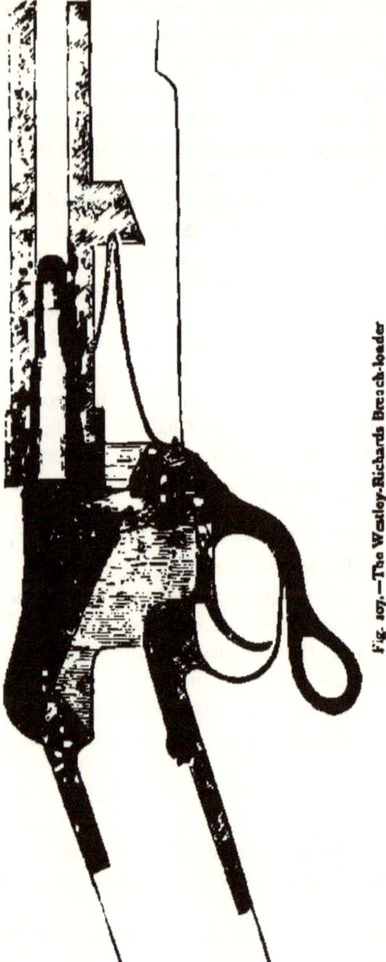

Fig. 107.—The Westley-Richards Breech-loader

action, which answers the purpose of a half-cock. The rifle can be loaded or unloaded with the trigger secured. The cartridge adopted for this rifle is the Berdan brass-drawn bottle-necked.

The bullet and rifling are upon Henry's plan. The parts are few and very strong, being only fourteen, and seven pins. This rifle performed well at the Wimbledon trials, 1870—see report.

We annex a representation of this rifle.

There is another Westley-Richards rifle, similar in

principle to the one described, but having a horizontal striker and hammer combined, that is acted upon by a main V-spring, which in this case is placed in the rear of the striker, thereby rendering the works more compact. By this means two or three parts are dispensed with, and the process of manufacture is made cheaper and more simple.

THE HENRY BREECH-LOADING RIFLE.

This rifle resembles the Sharp carbine. The breech is closed by a sliding vertical block, which is depressed for the admission of the cartridge by a lever underneath the trigger-guard. The piston passes diagonally downwards

Fig. 108.—The Henry Breech-loader.

through the breech block, and is struck by the hammer. The extractor is worked by the withdrawal of the breech block. This rifle has been fired twenty rounds in one minute and seven seconds. It won the £100 prize of the National Rifle Association as early as 1865. It also was highly spoken of by the Government Prize Committee. In their report, Feb. 2, 1868, they recommended the Government to award Mr. Henry £600; but they preferred the Martini action on account of the absence of a side lock. Mr. Henry has recently improved this breech-loader by making it self-cocking, and fitting an extractor similar to the Martini. He retains the same breech block, but entirely dispenses with the side lock. This reduces the movements to the same number as the Martini. Mr. Henry claims an advantage in being able to clean the barrel of his rifle from the breech-end. The barrel can also be examined more readily. This is an important point in all arms of precision. Nothing destroys good shooting qualities more than an accumulation of rust in the barrel. The Henry pattern of rifling requires great attention in order to keep it clean, and a frequent use of the wire brush is necessary.

The Martini rifle will not admit of the cleaning rod being inserted at the breech-end. In place of the rod they substitute a short twisted wire cleaner, which cleans the chamber and a few inches of the barrel only. The Henry rifle, in its improved form, fulfils the conditions of the Select

Fig. 109.—The Burton Breech-loader (No. 1).

Committee, and is preferred by many practical men to the Martini system.

THE BURTON BREECH-LOADER (NO. 1).

This rifle is upon the breech block principle. The block is hinged forwards and works downwards by means of a lever in front of the trigger-guard. A central-fire piston passes through the breech block. Its return is independent of a spring, being effected by the action of opening the breech, the same action also operating to extract the cartridge case. The ordinary lock and hammer are used. It can be fired about thirteen times a minute. It takes the Boxer ammunition.

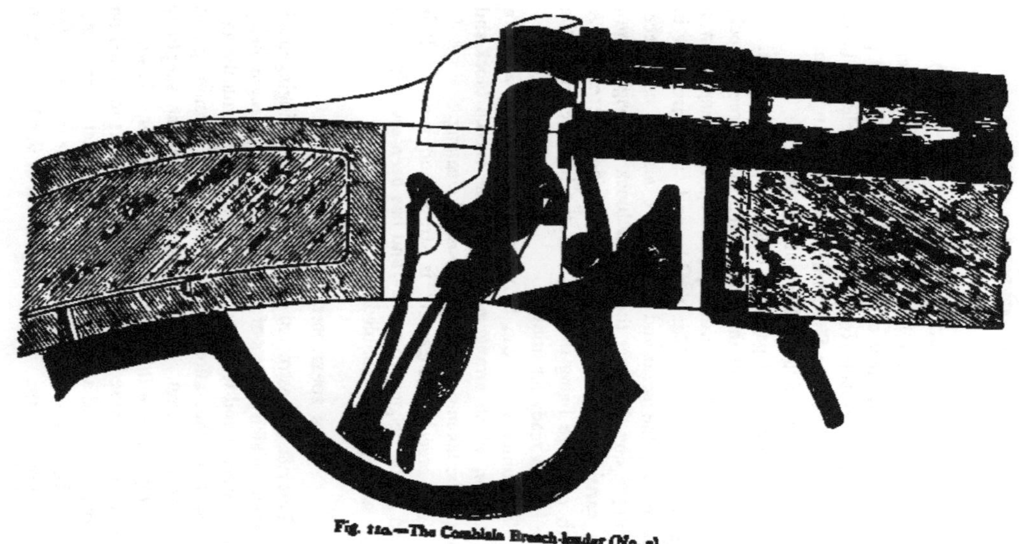

Fig. 110.—The Comblain Breech-loader (No. 2).

THE COMBLAIN BREECH-LOADING RIFLE (NO. 2).

This rifle is called No. 2 to distinguish it from the first Comblain, which is a modification of the Snider principle. The Comblain No. 2 has the vertical sliding block and guard-lever of the Sharp rifle; but the arrangement for exploding the cartridge is different.

The illustration at page 213 (Fig. 110) represents it.

The mechanism of the lock is fixed in the breech block, which consists of the ordinary main-spring acting upon a tumbler by a swivel. The tumbler and striker are made in one piece; the scear and trigger are also in one piece. By depressing the lever the breech block is brought down, the cartridge-case extracted, and the rifle is cocked. A fresh cartridge being inserted, and the lever returned, the rifle is then ready for firing.

The hinge screw can be removed without the aid of a turnscrew, which arrangement allows the breech block and lock to be taken out for the purpose of cleaning.

The breech arrangement is strong and simple. It is used by the Belgian volunteers, and has been severely tested both at Liege and Wimbledon.

THE SOPER BREECH-LOADING RIFLE.

This rifle is constructed on the side-hinged swinging-block principle, which admits of a cartridge of any length being truly inserted and retracted with its axis in line with the axis of the bore, thus affording great facility for examining and cleaning the barrel. It also combines extreme simplicity in manipulation with perfect safety to the user. It will be seen by the annexed illustration that a lever is mounted on the right side, and so conveniently situated that it can immediately be depressed after the discharge without altering the position of the arm. On the depression of this lever, which moves only 55 degrees in a circular direction, the striker is forced backwards, the breech-block raised, the hammer placed at full cock, and the empty case ejected. The ease with which this operation can be performed, at once explains the marvellous rapidity of fire that has been attained with this rifle; and it will be seen that this simple movement acts directly in four different ways, operating effectually on the striker, breech block, cock, and extractor, without the aid of spiral or any

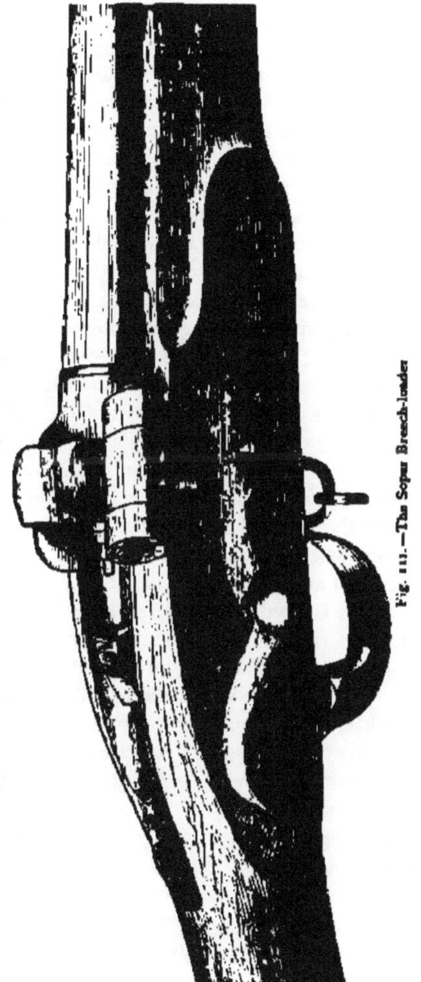

Fig. 111.—The Soper Breech-loader

other springs, except the ordinary main-spring taken from an Enfield rifle, and one flat trigger spring.

Although the mechanism of this rifle is extremely simple, yet on examination it will readily be seen that the hammer can in no case explode the cartridge until the breech-block is in its proper place. Experiments have been made to test the strength of the breech arrangements, in which, on several occasions, 200 grains of powder and a plug of lead 530 grains in weight have been fired. Full service cartridges have also been fired, having the base cut open with a saw, so that the gas has had a free passage into the breech chamber, without in any way interfering with the action of the working parts. It was also severely tried by placing it under water for a fortnight and then leaving it for the next fortnight exposed to the action of the weather in the open air, during the month of November, when, without oiling or cleaning in any way, it was repeatedly fired without miss or hitch of any kind occurring, the same rifle having previously successfully passed through the ordinary sand test, as used by the Select Committee at Woolwich. On reference to the sectional engraving (Fig. 112), it will be seen that the cock D, which is acted upon by an ordinary swivel and main-spring H, when released by the scear or trigger F delivers a blow upon the striker G directly in line with the axis of the barrel.

The striker G passes through the centre of the breech block A, and is secured and supported in its chamber by an ordinary military nipple I, which entirely prevents the striker from being driven backwards beyond its proper position, and at the same time forms a convenient fixing point for the ordinary snap cap, an arrangement that will be found of the utmost importance, when it is remembered that by far the greatest amount of work required of a military rifle is the snicking or snapping, which must be done under the present system of musketry instruction. The extractor E is fitted to slide in a groove at the top of the lock plate, and is acted on by a lever which acquires motion from the cock in such a manner that the old cartridge-case is slowly but powerfully drawn from the chamber, and afterwards rapidly ejected from the rifle. The stock is in one piece, and exceedingly strong.

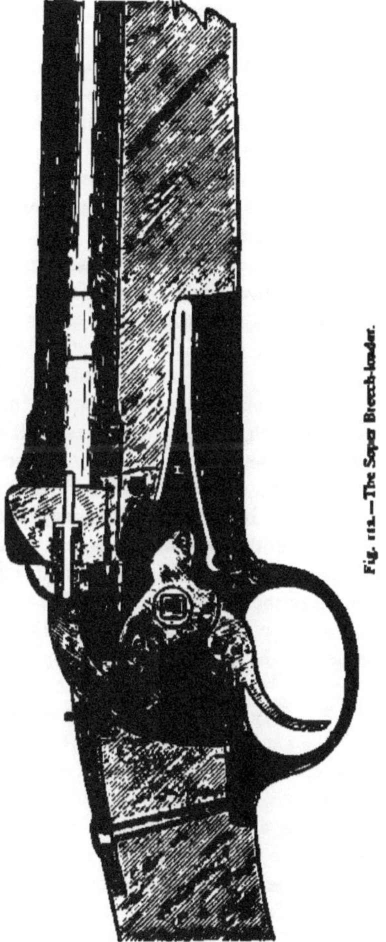

Fig. 112.—The Soper Breech-loader.

This rifle was first publicly fired for rapidity at the Basingstoke Exhibition, under the superintendence of the president, W. Spencer Portal, Esq., when it was actually loaded and fired sixty times in one minute, by Private Warrick, of the 1st Berks Volunteers, in the presence of a number of military officers, the time being kept by Major Grimstone, R.A.

This rifle was also tried on another occasion for accuracy and rapidity combined, when sixty shots were fired in two minutes, fifty-eight of which struck a target.

It also performed well at Wimbledon in July, 1870. We give a diagram of the shooting of this rifle in another chapter. The Henry barrel and ammunition are adapted for the Soper rifle.

The form of lever is well adapted for a military rifle, being so conveniently situated for the thumb, that it can be used well in a lying posture. There is less exertion required to load with this rifle than any we have yet seen, and the Government would do well to give it a fair trial by placing a number in the hands of the troops side by side with the Martini, Westley-Richards, and Henry (improved breech systems). All these rifles take the Government cartridge, and are rifled upon the Henry system. It is the opinion of many competent judges that the lever of the Martini rifle is most unsuitable for a military weapon. Should the bayonet still be retained in the service, this lever spoils the arm for a pike-handle. These points, and others of less importance, would be most satisfactorily settled by the Government adopting the above course.

THE AUSTRIAN WERNDL BREECH-LOADER.

The annexed illustration represents this breech-loader; it is constructed on the block system. The block works on an axis fixed below the bore of the barrel. Fig. 113 shows the breech action open to receive the cartridge, by turning the block to the left, which is accomplished by the thumb: the breech is then closed, and the piece is ready for firing. The cartridge extractor is acted on by the breech block. The cartridge is brass-drawn, central-fire, and is exploded by the ordinary lock and hammer. It resembles that of the Remington rifle; the bore is the same, viz., 450. The bullet is flat at the point; and is supposed to be more

Fig. 113.—The Austrian Werndl Breech-loader.

Fig. 114.—The Werndl Cartridge.

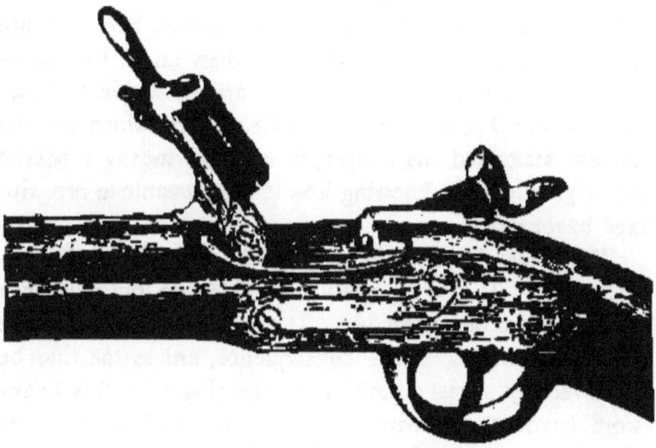

Fig. 115.—The Austrian Werndl Rifle.

accurate on that account. The Remington rifle and a converted muzzle-loader, much resembling the Braendlin-Albini (see illustration, Fig. 94, page 179), are used in the Austrian service.

THE BAYONET.

This arm is entirely of French origin, having been invented at Bayonne about 1644, and subsequently introduced into general use. It is an important weapon, much used by the French and English; but the Austrians, we are told, do not like it.

When first invented the bayonet was made to screw into the muzzle of the barrel. Grose, in his "Antiquities," bears testimony to the French origin of the bayonet and its improvements in the following anecdote, in which the English received a practical lesson they were not likely to forget:—

In one of the campaigns of William III. in Flanders, occurred an engagement in which fought three French regiments, whose bayonets were made to fix after a new fashion. One of these advanced against the 25th regiment with fixed bayonets. Colonel Maxwell, who commanded it, immediately ordered his men to screw their bayonets into their muskets, the only mode they then knew, to receive them. Great was his surprise when, arrived within a proper distance, the French threw in a heavy fire which for the moment staggered his people, who by no means expected such a greeting, not knowing how it was possible to fire with fixed bayonets.

The sword bayonet is very much in favour at the present period. It may be useful as a side arm, but fixed to a rifle it is most objectionable. It spoils the balance of the arm; bad shooting is the consequence, unless the rifle be supported by a rest; besides, in charging with this heavy sword bayonet, the barrel of the rifle would get bent or otherwise injured, and rendered useless as an arm of precision. It is the opinion of many military men that the bayonet will be shortly discarded, but we do not think that will happen for some time to come.

Instead of dispensing entirely with the bayonet, a more convenient arrangement should be adopted. We annex an illustration representing our improved bayonet, which we think will be found to answer every purpose.

As the forend of the Martini rifle stock is of little service except to protect the barrel, we make use of it for a bayonet-case. This bayonet can then be pushed forward and secured instantly when required, and, except when wanted for use, it is kept in the stock like a pencil in a case.

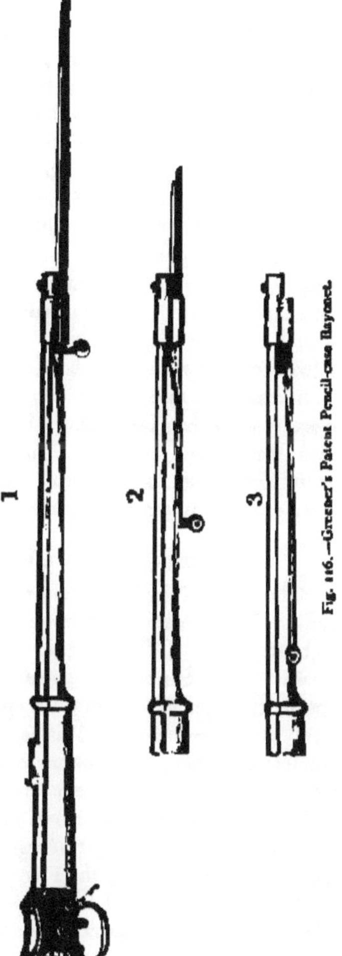

Fig. 116.—Greener's Patent Pencil-case Bayonet.

When it is fixed, it is secured by a kind of nose-cap, which forms the end of stock, passes round the barrel, and is attached by soft soldering or by a screw-pin. The handle is a knob which projects from the stock, and slides up the ramrod groove; it is secured by turning it on one side into a notch in the nose-cap, with the addition of a spring catch. When the bayonet is brought back into the stock, the knob is turned down flush out of the way. The ramrod is fitted on the side of the stock in this arrangement. There are several advantages in this plan—no bayonet-scabbard or frog, which very much incommodes the soldier in skirmishing, is required; the rifle can still be made the regulation weight with the bayonet included, this would enable the soldier to carry more cartridges, or a light spade, which will be necessary in modern warfare; this bayonet can be fixed more quickly than the ordinary kind, can be made stronger than the triangle bayonet, would resist a greater strain, as it lies closer to the barrel, and would be altogether more convenient, especially when skirmishing or in rifle-pits.

THE SHOOTING OF MILITARY BREECH-LOADERS.

The range of the Prussian needle-gun is from 600 to 700 yards. It possesses a moderate degree of accuracy with a high trajectory. The bore is about 16; the bullet is smaller, being 534 and 2 diameters long. The charge of powder is 70 grains. It can fire upon an average seven or eight shots per minute. This is considered now a very indifferent weapon, when compared with the latest improved arms; but with all its imperfections it has proved a most destructive weapon in the hands of the Prussian soldiery, who have exhibited great coolness and deliberation in all their engagements during the Franco-German war.

The extreme range of the French Chassepot is given as 1,800 yards. This great range is obtained by reducing the bore and the weight of the bullet, which is only 380 grains, and using the large charge of 85 grains of powder. The bore of the barrel is smaller than the Henry rifle. Although the range of this rifle is superior to the needle-gun, it is not so accurate. It can be fired about ten shots per minute.

The range of the Snider is about 1,300 yards. It possesses a greater degree of accuracy than the Chassepot, although the bore is much larger, being 577. The bullet weighs 530 grains, the charge of powder 75 grains. It can be fired by troops from twelve to fifteen times per minute. The trajectory is, however, very high, being at 500 yards 11 feet 10¾ inches. Although only a converted arm, it is a really good and serviceable weapon, and will bear comparison with the new arms of other European nations. Flat trajectory is one of the essential points in a military rifle, and to obtain this it is requisite to reduce the bore and increase the charge of powder. This was fully worked out by Mr. J. Whitworth, in 1857, by desire of the Government; the result being the adoption of a bore ·450, and a bullet 3 diameters long. That the proportions adopted were right is shown by the fact which the Committee reported, viz., that the makers of every small-bore rifle, having any pretensions to accuracy, have copied to the letter the three main elements of success adopted by Mr. Whitworth, viz., diameter of bore, degree of spiral, and large proportion of rifling surface. The Henry breech-loader is of the Whitworth bore, but the spiral is one turn in 22 inches. The bullet is shorter and lighter, but the principle is essentially the same. The form of rifling is an improvement; the grooves are not so deep. (See the illustration, Fig. 62.)

The Henry barrel and bullet have been proved to fulfil all the requirements for long range and accurate shooting. The trajectory is very flat. The barrel has no tendency to lead or foul. A great number of shots can be fired without cleaning. We give the ascertained trajectory of the Snider, Chassepot, and Martini-Henry. The Snider, at 500 yards with regulation charge, is 11 feet 10¾ inches at the highest point; the Chassepot is 10 feet at the highest point; the Martini-Henry is 8 feet 1¾ inch at the highest point. It is well known that a bullet in its flight describes a curve: thus the necessity for sights which give elevation. For example, the Snider bullets would rise above the line of aim 11 feet 10¾ inches at the highest point, returning again to the line of aim 500 yards from the muzzle. The Martini-Henry would rise only 8 feet 1¾ inch above the line of aim; in this consists the chief superiority of the Martini-Henry.

Without the use of sights upon a rifle, long range cannot be obtained with certainty; without sights the bullet would strike the ground at one-fourth the actual range. The intention of sights is to elevate the muzzle imperceptibly.

For a better illustration, the safe distance for infantry opposed to the Snider with the 500 yards sight up would be between 92 and 438 yards; with the Martini-Henry, the safe distance would be between 135 and 396 yards. Now, taking the height of the soldier at 6 feet, if the trajectory could be so flattened as to ensure, at a range of 500 yards, that the highest point of the curve would not exceed the height of the soldier, it is evident that any error in judging distance, in the hurry of file or rapid firing, would make but little difference in effect on a line, much less on a column.

It is an easy matter to flatten the trajectory of a rifle, by reducing the weight of the projectile and increasing the charge of powder, for any range under 600 yards; but beyond that distance the bullet would fall rapidly, and the long range would be sacrificed.

It is impossible to combine in a rifle of a moderate weight very flat trajectory and long range.

"It may be interesting at the present time to recall the results of a trial which was made at Woolwich last year between the Chassepot and the Martini-Henry rifle, and which are recorded in the proceedings of the Royal Artillery Institution. Captain Simon, of the French Artillery, submitted three improved Chassepots. They were fired in comparison with the Martini-Henry at 500 yards range; the following are the figures :—

Arm.	By whom Fired.	Mean Deviation of Twenty Shots.
Martini-Henry ...	Sergt. Bott, R.M., kneeling position ...	1·03 feet.
Chassepot	Captain Simon, from shoulder-rest ...	2·78 „
Martini-Henry ...	Captain Simon, from shoulder-rest ...	0·97 „
Chassepot	Sergt. Bott, R.M., kneeling position ...	3·02 „
Martini-Henry ...	Edward Ross, Esq., sitting position ...	0·96 „
Chassepot	Captain Simon, from shoulder-rest ...	2·78 „
Martini-Henry ...	Captain Simon, from shoulder-rest ...	1·62 „
Chassepot	Captain Mackinnon, sitting position ...	2·38 „

Thus the accuracy of the Martini-Henry far excelled that of the Chassepot. We may also add that the accuracy of the Chassepot, as here exhibited, was far inferior to that of the Snider rifle."

It will be seen by the above report that the shooting of the Chassepot was very bad for a small-bore; this is owing to the form of rifling, which is the very worst pattern that could have been selected for a breech-loading arm. The grooves are too few and the lands far too wide. The bullet being larger than the bore, it must be driven through the barrel. One-half the surface of the bullet is consequently cut away by the broad lands. As we have before stated, the bullet leaves the barrel nearly four square. The best form of rifling for a bullet like the Chassepot is polygroove, with very narrow lands. (See our illustrations of rifling.)

The French authorities have discovered that the Chassepot rifle is a mistake. They intend to adapt it for the metallic cartridge at an early period.

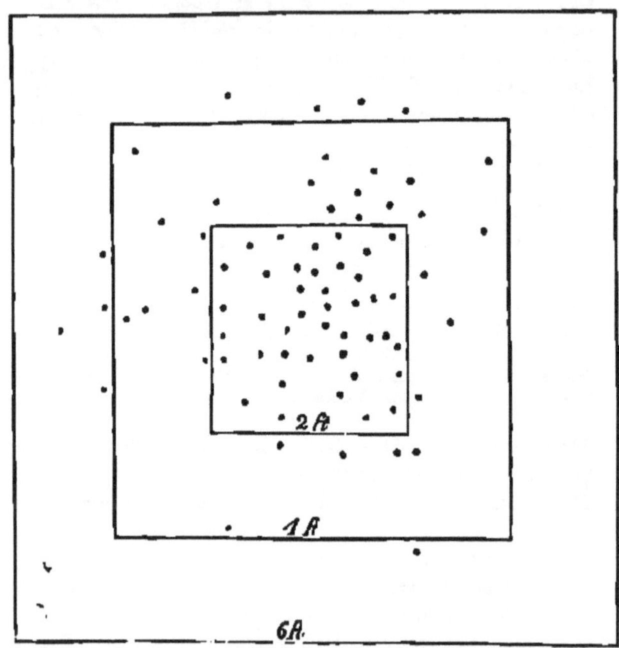

Fig. 117.—Target of the Soper-Henry Rifle at 500 yards.

SHOOTING AT WIMBLEDON, 1870.

The performances of the Government Martini-Henry rifle at Wimbledon were such as must have given satisfaction to every one except the inventors of rival weapons. In a year of exceptionally fine and remarkable breech-loading shooting, the Martini-Henry has held a high, if not the highest place. Perhaps, taken all round, the rifle did better than any other. The extraordinary shooting of Mr. Farquharson, which is quite without a parallel in the history of breech-loaders, gave the Henry a better position for extreme rapidity and accuracy. At a short range the Soper-Henry, an amazingly quick rifle, held second place.

At 500 yards the Martini-Henry was second, third, and fourth. At 800 yards it took the whole of the prizes given at that range, including the Duke of Cambridge's cup.

We give a diagram of the shooting of the Soper-Henry at 500 yards with the regular military sight.

Better shooting can be made with this rifle if used with the improved sights. The Henry rifle has been tested at Woolwich for accuracy at 1,200 yards from a machine-rest, and with all the improved appliances. It has put all its shots in a square target 2½ feet. This is much better shooting than can be obtained from the shoulder at 500 yards.

EXPERIMENTS TO ASCERTAIN THE FALL OF PROJECTILES WITH MILITARY RIFLES.

The first experiment was with the Enfield Rifle, 1853 pattern, from a shoulder-rest; the rifle and point aimed at were 5 feet above the level of the ground. With the sight for 100 yards, and regulation cartridge, the mean of ten shots struck the ground at 277½ yards.

The Whitworth rifle, 2-ft. 9-in. barrel, 480-grain bullet and 75 grains of powder, struck the ground at 321½ yards.

The 1853 pattern rifle was again tried, fixed in the machine-rest, perfectly horizontal, "no elevation," 4 feet 10 inches from the ground. The mean of ten shots struck the ground at 188½ yards.

A Snider long Enfield, laid perfectly horizontal in the machine-rest, 3 feet high; the mean of ten shots struck the ground at 163 yards.

The fall of the bullet from a long Enfield Snider at 100 yards is 12 inches; at 125 yards, 18½ inches; at 150 yards, 26⅝ inches; at 175 yards, 35⅝ inches; at 200 yards, 51¾ inches.

We may remark that the fall of a bullet fired from a shoulder-rest appears to be less than when fired from a machine-rest; this is accounted for by the fact of a sight being used for 100 yards, and a slight tendency of the muzzle to rise when fired from the shoulder, there being no elevation or possible rise at the muzzle when fired from the machine-rest.

The trajectory of a long Enfield Snider, 577-bore, at 500 yards. The height of the bullet above the *level of axis* of bore is as follows:—

Range in yards	50	100	150	200	250	300	350	400	450	500
Height in feet	3 8½	6 6	8 10¾	10 9	11 5½	11 5½	10 4½	8 2½	4 10½	—

The same rifle when fired at 400 yards:—

Range in yards	50	100	150	200	250	300	350	400
Height in feet	2 0½	3 9½	5 4½	6 9½	6 7½	5 11½	4 9½	...

We give the accompanying illustration to show the difference between the trajectory of the Snider and the Martini-Henry. The dotted line represents the flight of the Henry bullet, 450-bore; the other line is the flight of the Snider bullet, 577-bore, which will strike the ground short of 300 yards, aiming, of course, as represented. To hit the mark at 300 yards it would require considerable elevation, and would then fly over the head of the man at 100 yards.

Recoil.—1853 pattern rifle, service ammunition, 50½ lbs.; long Enfield Snider, Boxer ammunition, 47½ lbs.; Whitworth, hexagonal, 530-gr. bullet, 85 grs. powder, 54¾ lbs.; Whitworth, cylinder, 530-gr. bullet, 70 grs. powder, 47½ lbs.;

Westley-Richards, cavalry, breech-loader, 43¾ lbs.; cavalry carbine, 577-bore, 43½ lbs.; cavalry carbine, with 2 drams powder and 480-gr. bullet, 39 lbs.

RAPIDITY OF FIRE WITH BREECH-LOADERS.

Very rapid firing in actual warfare is not at all desirable. Ten well-directed shots per minute would prove much more effective than thirty fired wild; but the rifle that can be fired the greatest number of rounds per minute must certainly be the easiest to load, and consequently the most desirable weapon to place in the hands of troops, as it would tend to give confidence and coolness to the men. Some persons have objected to breech-loaders on the ground that they would only tend to waste of ammunition; but this theory has been entirely upset by the Prussian campaign of 1866. It was found that the quantity of shots fired by the Prussians did not amount to five shots per man; and those soldiers who were engaged in the thickest of the fight had only used about one-half the number of cartridges allotted to each man.

The following interesting account, which we take from the *Globe*, as being contributed by Lieut.-Colonel Reilly, C.B., Royal Horse Artillery, gives details of the expenditure of

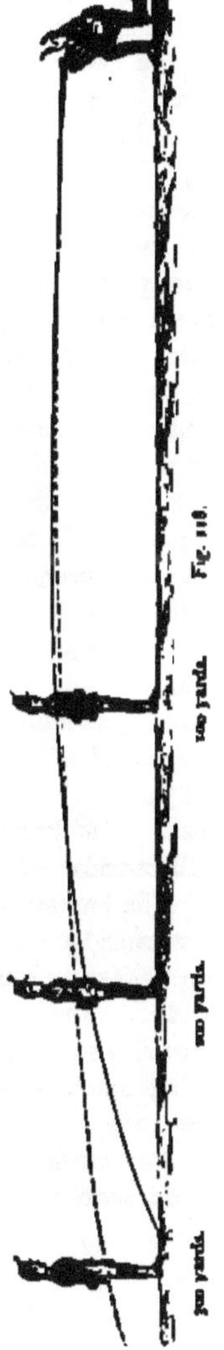

Fig. 118.

small-arm ammunition in the Prussian army. He says, "The greatest number of rounds fired during the war in the Second Army was by the 43rd Regiment at Trantenau—43 rounds per man. At Königgrätz the first battalion of the Guards, 915 strong, fired 12,694 rounds, or 13·8 rounds per man; the third battalion of the same regiment, 901 strong, fired 12,250 rounds, or 13·3 rounds per man. The expenditure of the 5th Corps, about 20,000 strong, in the actions prior to Königgrätz, where it did not fire a shot, was 517,000 rounds, or about 26 rounds per man. In the First Army, the 27th Regiment, 2,550 men, at the battle of Königgrätz, expended 30,000 rounds, or under 12 rounds per man. In some cases the First Army drew upon the regimental reserves. Thus, the three battalions of the 71 regiments at the battle of Königgrätz expended 184,000 rounds, or about 72 rounds per man; it must, however, be borne in mind that the First Army was in action four hours longer than the Second. There was a total of about 110,000,000 cartridges prepared on the Prussian side for the war of 1866, and the materials for 70,000,000 more were ready. The total expenditure during the war of all the armies, of about 400,000 men, was only 1,854,000 rounds, or between four and five rounds per man. The Prussian infantry soldier carried 60 rounds of ball cartridge—40 in his pouch, and 20 in his knapsack."

The improvements which have been made in the needle-gun cartridge reduce the weight of the bullet considerably; we believe 92 rounds can now be carried by the Prussian soldier. The same number of Chassepot cartridges can be carried also; but the Snider and Martini-Henry cartridges being much heavier, 60 rounds only can be carried conveniently.

For rapidity the Soper rifle stands unrivalled, as many as 60 shots per minute having been fired out of it by an expert. When aiming at a target it is possible to fire 30 shots per minute; 60 shots have been fired in two minutes, 58 of which struck the target. With such rapid firing as this, the barrel becomes so hot that the back-sights fall off, they being only attached by soft solder; but this can be prevented by screwing them on. Another inconvenience resulting from rapid firing is, that the hand would be burnt unless protection was provided.

Westley-Richards has met this difficulty in his own rifle by surrounding the barrel with wood at the breech-end, which is a simple and effective plan.

The Henry, Westley-Richards, and Martini rifles are all that can be desired for rapidity; but the Chassepot, with all the advantages of the metallic cartridge, can never be a quick weapon, as the cocking arrangement is very tedious, on account of the force required by the thumb to bring the piece to full-cock. The easier a breech action can be manipulated, the better will be the shooting, especially when it is continuous. We give a scale for rapidity of each weapon without taking aim, which may be considered a fair average:

The Soper	50 per minute.
The Martini	40 " "
The Westley-Richards	38 " "
The Henry	38 " "
The Remington	30 " "
The Chassepot	19 " "
The Snider	18 " "
The Braendlin-Albini	18 " "
The Berdan, Russian	18 " "
The Needle-gun	9 " "

CARTRIDGES FOR MILITARY BREECH-LOADING SMALL ARMS.

Metallic cartridge-cases or chambers for breech-loading arms are not at all a novelty. We have many specimens in our museums of metallic chambers for loading jingalls and other arms at the breech—proving that they were known to Oriental nations long ago. We must therefore consider them only as reproductions in an improved form, and adapted to breech-loaders of the present day. The real secret of success in these metallic cartridges is the application of the percussion principle. The Lefaucheux cartridge (1836) was most unsuitable for a military rifle, on account of the projecting pin. About 1860, the Americans adopted the rim-fire copper cartridge, which was used successfully in the Spencer rifle. The Pottet cartridge appears to have been introduced into this country about this time, by Mr. Daw, of London, and is known as Daw's central-fire cartridge.

This is decidedly the best form of cartridge invented; the

Section of "Boxer" Cartridge, ·577 Bore.

Fig. 119.

mode of ignition is certain, and the arrangement of the cap is most convenient and safe. Col. Boxer's patent cartridge is an improvement upon the Daw; the base and mode of ignition are precisely the same, but for the pasteboard tube he substitutes one of brass, formed by coiling sheet metal in double folds and covering it with waterproof paper. It will be seen, by examining the annexed illustration, that there is a papier-mâché wad which forms the base and contains a metallic dome, which receives the cap and anvil. The first Boxer cartridges made had brass rims, resembling the cartridge used in sporting guns.

The improved plan is to form the rim by attaching an iron disc to the base of the cartridge, and by rivetting it to this dome. This plan insures the rims being made all of one uniform thickness, which is very essential in a military rifle.

Colonel Boxer obtained a patent for his cartridge (15th January, 1866), which he designed specially for the Snider rifle. One of the good points in it is that the case can be easily extracted from the barrel after firing. The cartridge is made much smaller than the bore of the chamber, in order to admit it freely; on explosion the case expands considerably, and fills up the chamber, effectually preventing the escape of gas.

This form of case is used for the Martini and many other systems of breech-loading rifles, with great success. It has

Fig. 122.

lately been improved by Messrs. Eley Bros., who have taken out a patent for making it bottle-necked. This neck, or smaller diameter of the cartridge-case, is formed by means of a die, so arranged as to form flutes or folds in that part of the case, and consequently reduce the diameter; the paper covering is dispensed with, a paper lining being substituted. The weight of the case is 165 grains; the length of cartridge complete, 450-bore, is 3 inches, with the Henry bullet, the charge of powder being 85 grains, the weight of bullet 480. We give an illustration of this cartridge, full size.

As we have before stated, Messrs. Eley Brothers have recently introduced a bottle-necked cartridge-case made of sheet tin; this is much stronger than brass, and can be made much cheaper. We consider it to be far superior to the brass Boxer, and believe it will be ultimately adopted by the Government for the regulation cartridge.

The Daw military cartridge is similar to the Boxer; the brass tube is of one thickness, and soldered up the side. This cartridge was submitted for trial, with about forty-nine others, to the Select Committee, who awarded Mr. Daw £400 for the invention.

THE SOLID BRASS-DRAWN CARTRIDGE.

These cartridges were first made in the United States. The first manufactured contained the ignition in the rim; but this mode proved upon trial to be unsatisfactory, on account of the frequency of missfires; they were also liable to split and expand at the base. They are now superseded by the central-fire, which have a solid base the thickness of the rim. All the early cartridge-cases were made of Lake Superior copper; but it has since been found advisable to substitute brass, on account of its elasticity.

The Berdan cartridge-case is most extensively used in the American arms, and is the invention of General Berdan, of the United States Army. It is made in the following manner:—punched from sheet metal and drawn to the required length in six operations, headed, chambered, anvil drawn up, reinforced, pierced, and necked in five more. The shell is reinforced at the base by a ring or perforated saucer swaged in, which guards against the possible escape of gas through cracks, or invisible defects of any kind in the flange, a not unfrequent occurrence with the ordinary drawn metallic cartridge. The case is necked to suit the calibre of the projectile. The case is primed externally, by the insertion of a shallow, saucer-shaped percussion cap, which, on being pressed into the cap chamber, rests on the metal, so as to insure that the head of the cap containing the fulminate does not touch the anvil, thus carrying out in a perfect manner the idea of the safety shoulder-anvil. When the cap is home in its place, the head being well inside the level of the base of the cartridge insures safety from ignition, except by a blow from the striker, which indents the cap. We give an illustration showing this cartridge cut open.

This cartridge case, being much thinner at the base than the Boxer, will admit of being made shorter and will contain the same charge of powder. The principal advantage of this cartridge is, that it can be recapped and fired a great number of times. It is made smaller than the chamber of the barrel, to allow for expansion. These cartridges have been improved and successfully manufactured by Messrs. Ludlow, of Birmingham.

Fig. 121.—The Berdan Cartridge.

A short, solid, cannelure bullet, lubricated with wax, is used in cartridges for American arms. The Henry bullet, wrapped in paper, will give far better shooting at long ranges.

The primer used in the Boxer cartridge we consider far more certain of ignition than the Berdan, on account of the detonating powder being more concentrated, and the form of anvil being better adapted for the purpose. The numerous trials made by the Government have proved that the Boxer cartridge approaches as near perfection as possible. From the reports of the Committee we find that the Martini rifle, with the improved spiral spring, fired 26,463 rounds of ball and blank ammunition with only nine missfires, being a percentage of 0·34.

PERCUSSION CAPS AND CARTRIDGES NOT DANGEROUS.

An impression, entirely without foundation, has extensively prevailed that percussion caps were dangerous articles. In case they should be accidentally exposed to fire or concussion, it was thought that they might be exploded in bulk; and that, therefore, packages containing percussion caps in transit, or when stored, would be a cause of danger. The railway companies, under this impression, had recently decided to treat these articles as dangerous, and had placed important difficulties in the way of their carriage. This alarm, too, was extending to the ports, and serious obstacles were beginning to be placed in the way of their shipment.

These fears were clearly without any basis, and with the object of removing them a series of experiments were exhibited in Birmingham, on the 28th of April, at the instance of the Chamber of Commerce, at which a considerable number of gentlemen were present, including delegates from the principal railway, canal, marine fire insurance companies, and other important bodies.

These experiments demonstrated that if one cap is ignited by concussion or fire, the fire is not communicated to other caps in its close neighbourhood, and, therefore, percussion caps cannot be exploded in bulk.

The first four tests were intended to show that in the event of fire occurring from collision or otherwise, percussion caps as packed for transit, whether in small or large quantities, would not explode, or increase the danger or damage caused by an ordinary fire.

The third experiment was made at the muffle used in the second experiment. In this experiment 50,000 percussion caps were used. These were contained in ten paper parcels, and both the packing-box and its contents were precisely such as are sent out by the manufacturer to the merchant. The package was placed in the red-hot muffle. In a few seconds it was enveloped and concealed by flames. After burning a little over three minutes there was no more flame, and the packing-box could again be seen, charred and red-hot. The whole of the percussion powder must have been burned out of the caps long before this time, but no explosion or noise of any kind was observed. On revisiting the muffle half-an-hour afterwards, it was found that the whole of the packing-box had burned away; the upper part of the red-hot mass had fallen, but the lower part consisted of the tin boxes still piled upon one another, undisturbed by the action of the fire or the percussion caps contained in them.

The experiments eight and nine were then made at the railway station. The bags of percussion caps were hung on a solid buffer, and struck by a locomotive weighing 45 tons, with a blow of at least 200 tons, and those caps only exploded which fell on the rails and were subsequently passed over by the wheels of the engine and tender. Bags containing 20,000 caps were placed on the rails, and the locomotive ran over them at a high speed. Those caps only exploded which the second pair of wheels struck, and the explosion did not spread to any of the caps immediately off the rails, although out of the same bag.

These experiments proved so satisfactory to the railway authorities and some of the shipping companies, that they will convey these caps and cartridges now without hesitation, which is a great convenience to the makers, merchants, and consumers.

THE AMERICAN GATLING GUN.

The mitrailleuse is not altogether a modern invention, for it is mentioned in Grose's "Military Antiquities," 1801, that in England, in the year 1625, under Charles I., a patent for a new invention was granted to a certain William Drummond. The invention is described as a machine composed of a number of muskets joined together, by the help of which two soldiers can oppose one hundred, and named, on account of its effect, Thunder Carriage, or, more usually, Fire Carriage.

The Gatling Gun is an American invention; we give an illustration representing it.

Fig. [illegible].—The Gatling Gun.

There are ten barrels bound together, revolving round a central axis parallel to their bore by means of a hand crank. The cartridges are placed in a hopper at the left side of the gun. As each barrel comes opposite this hopper, a cartridge falls into a groove, and is pushed into the breech gradually by a plunger. The barrels are discharged when they reach the right side of the piece. The cartridge cases are extracted by an extractor fixed to the plunger. This plunger retracts immediately after the discharge, bringing out the empty cartridge-case, or a loaded one should it have missed fire.

Every barrel has a separate lock, which is connected with the exploding piston, that passes through the breech plunger.

There is an additional handle to give a lateral motion to the gun. The cartridge is a central-fire, brass-drawn, on the Berdan principle.

The United States Government and Russia have adopted the Gatling—France and Austria, the Mitrailleuse.

Two of these Gatling-Mitrailleuse were fired at Shoeburyness. The first, having a calibre of one inch, and throwing a ball weighing 8 oz., was fired at a target 36 feet broad and 9 feet high, at a distance of 2,100 yards. Owing to a couple of cartridges placing themselves athwart the loading machinery, only 121 shots were delivered in two minutes, resulting in 12 hits.

A smaller Gatling, calibre 65, throwing a 3-oz. ball, was tried. It despatched 308 bullets and produced 45 hits. The larger Gatling was then pointed at three targets, each of the same size as the last, but placed in rows to represent a column of troops. 238 shots were fired and about 90 hits were scored.

The smaller Gatling fired at the column 348 bullets in two minutes, making 165 hits at 2,100 yards. These results are far beyond anything which our artillerists had expected from any mitrailleuse.

Hitherto these engines had done fairly at short ranges, but these performances would have been excellent for the best field-pieces, at the very fair range of 2,100 yards, or considerably over a mile.

The War Department are so satisfied with the Gatling that they have determined to order a battery of 60, which will be supplied as weapons of the service. The destructive power of this weapon is very great; for although it has a bore of only 0·42 inch, each battery (it is proved by trials) will strike, wound, or kill, amongst broken infantry, on uneven ground, at ranges from 200 to 1,000 yards, at the rate of 900 per minute; and in close column of infantry, a battery would hit 1,200 men per minute. The weapons just ordered, 300 or 400 in all, will be manufactured at Colt's Armoury, Hartford, United States.

THE MITRAILLEUSE.

We give a representation of the mitrailleuse, which has 37 barrels, all bound together like a faggot of sticks, and soldered fast in that position. They are open at both ends, and behind is a wrought-iron framework, to support the breech-loading apparatus. A breech block, containing a separate spiral spring and steel piston for each barrel, slides backwards and forwards behind the barrels, worked by a lever. When the breech block is drawn back, there is a space sufficient, between it and the barrels, to slip down vertically a plate pierced with holes containing cartridges, one for each barrel. Then the breech block is pressed forward by means of the lever, and this action both closes fast the backs of all the barrels, and compresses the spiral springs, so that they are ready to thrust their pistons forward suddenly against their corresponding cartridges, and so ignite them, but for a certain hindrance. This hindrance is a thin steel plate in front of the pistons, but is movable out of the way by the action of a handle. As the handle is turned fast or slow, the plate slides out of the way quickly or slowly in proportion, and permits either one piston after another to strike and discharge its cartridge at intervals of any duration, or, by rapid turning of the handle, all the pistons to strike their cartridges so rapidly that the 37 barrels are discharged almost simultaneously, or as nearly so as the rifles of a company of infantry ordered to fire a volley. The barrels being practically parallel, the bullets fly pretty close,

Fig. 123.—The Mitrailleuse.

and great destruction must occur if the piece be only properly laid on the object. As ten platefuls of cartridge or 370 bullets can be discharged in one minute, it is evident that nothing could pass a bridge or doorway, a narrow path, or the ditch of a fortress guarded by mitrailleuses well served and protected. The machine can be easily worked by two men, possibly even one; but it is too heavy to be conveyed otherwise than on a small carriage, and a carriage involves horses. It is not supposed that it can meet and master a field gun. It occupies a place between field artillery and infantry.

And here our labours end. It is, however, our intention to take careful note of the progress made in the art of constructing breech-loading fire-arms, the discoveries made from time to time in the science of gunnery, and the elaboration of practically valuable explosive compounds; but until we have gleaned and gathered together subject matter sufficient for another volume, we bid the reader adieu.

NOTES.

The Martini-Henry rifle is now adopted by the Government, to be used with the Boxer bottle-necked cartridge; weight of rifle reduced to 8¼ lbs. With the improved tumbler rest, as adopted, the trigger can be made to pull off at 3 lbs.

The tin Boxer cartridge is adopted for the Snider rifle. It might be also used advantageously with the Martini-Henry rifle.

ELEY'S AMMUNITION.

CENTRAL-FIRE CARTRIDGES

(as originally made for C. W. Lancaster in 1857)

OF CALIBRES 4, 8, 10, 12, 14, 16, 20, 24, ·442, ·410, ·360.

The Paper Cases of these Cartridges are rendered Waterproof, and, being primed with Eley's Double Waterproof Caps, are certain of ignition in all climates. They may be kept loaded for any period without deterioration.

Pin-fire Cartridges of Calibres 4, 8, 10, 12, 14, 16, 20, 24, ·28, ·500, ·410, and ·360.

These Cartridges are Waterproof, both in the Paper Cases and the Caps, under the same conditions as the Central-fire.

Eley's Double Waterproof Central-fire Caps are now well known and approved, being found superior to all others for their certainty and rapidity of fire, either in dry or wet weather.

For India and the Colonies, or any climate where Caps may be exposed to great vicissitudes of heat, cold, or moisture, they are particularly recommended, as they cannot be injured by any amount of exposure to wet or heat, nor their qualities impaired if kept for years in a tropical climate. The ignition at all times is safe and certain, whilst in humid weather the discharge is as instantaneous as with the ordinary Cap on the driest day. They have been much approved for the Rifle in Foreign Field Sports, where the Cap is often allowed to remain a long time on the nipple.

Being perfectly Waterproof, they will bear immersion in sea water for days without injury.

Waterproof Military Caps for Enfield and other Rifles.

Metal-lined Percussion Caps for Colt's and all other makers of Revolvers.

Thick Felt Wads, made expressly for Breech-loaders, for use over the powder; also **Cloth Wads**, as used over the shot.

Patent Shot Concentrators for Breech-loading Cartridges, the use of which in any Breech-loader gives an increase of from 20 to 30 per cent., both in pattern and penetration.

Chemically prepared Cloth **Gun Waddings**, to prevent the leading of guns, warranted not to blow to pieces in the barrel; and every description of **Sporting Ammunition** (Gunpowder excepted).

SOLE MAKERS OF THE

ELEY BOXER CARTRIDGE,

As adopted by H.M.'s War Department for the Snider and Martini-Henry Rifles.

ELEY BROTHERS,

254, GRAY'S INN ROAD, LONDON.

(WHOLESALE ONLY.

JAMES DIXON & SONS,

CORNISH PLACE, SHEFFIELD,

And 37, *LUDGATE HILL, LONDON,*

MANUFACTURERS OF THE BEST SPORTING APPARATUS FOR BOTH BREECH AND MUZZLE LOADING GUNS.

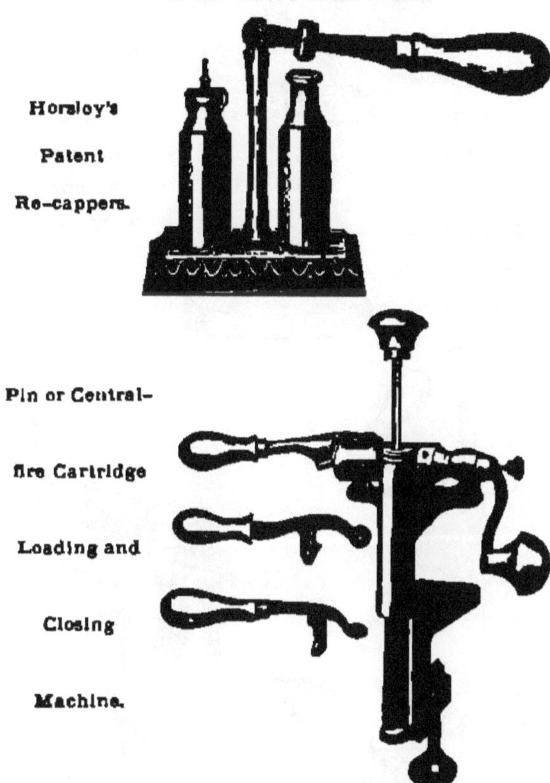

LIST OF WORKS

PUBLISHED BY

Messrs. Cassell, Petter, & Galpin.

Bibles and Religious Literature.

Bible, Cassell's Guinea. With 900 Illustrations, Family Register, and References. Royal 4to, cloth gilt, gilt edges, £1 1s.

Bible, Cassell's Illustrated Family. Toned Paper Edition. Half morocco, gilt edges, £2 10s.; full morocco antique, £3 10s.; best full morocco elegant, £3 15s.

*** Cassell's "Illustrated Family Bible" has attained a circulation of nearly *Half a Million*.

Bible, The Child's. 830 pp., 220 Illustrations. Demy 4to. Being a Selection from the Holy Bible, in the words of the Authorised Version. Cloth elegant, gilt edges, £1 1s.; flexible leather binding, hand tooled, gilt edges, £1 10s.; best morocco elegant or antique, £2 2s. Also kept with clasps, and in illuminated leather.

*** Nearly 60,000 copies of "The Child's Bible" have been already sold.

Bible, The Doré. Illustrated by GUSTAVE DORÉ. Complete in Two very handsome Volumes. Bound in cloth gilt, £8; morocco, gilt edges, £12; best polished morocco antique extra, £15.

Bible Dictionary, Cassell's. Complete in One or Two Volumes, strongly bound in cloth, 21s.; in One Volume, strongly bound in russia or morocco, 40s. 1,159 pp., and nearly 600 Illustrations. Imperial 8vo.

Commentary, Matthew Henry's. Complete in Three Volumes. Cloth, lettered, £2 12s. 6d. Demy 4to, 3,308 pp.; numerous Illustrations and Maps.

Family Prayer Book, Cassell's. Cloth, 7s. 6d.; with gilt edges, 9s.; morocco antique, 21s. 398 pp., demy 4to.

Quiver, The. Volume for 1870. 840 pp. letterpress, and numerous Engravings. Cloth, lettered, 7s. 6d.; gilt edges, 8s. 6d.

Children's Books.

Shilling Story Books. 96 pp., fcap. 8vo, cloth, lettered. With Illustration in each Volume.

Lottie's White Frock.
Helpful Nellie.
Only Just Once.
The Boot on the Wrong Foot.
Little Content. By EDITH WALFORD.
Little Lizzie. By MARY GILLIES.
Luke Barnicott. By WILLIAM HOWITT.
My First Cruise. By W. H. G. KINGSTON.
The Boat Club. By OLIVER OPTIC.
The Delft Jug. By SILVERPEN.
The Elchester College Boys. By Mrs. HENRY WOOD.
The Little Peacemaker. By MARY HOWITT.
Jonas on a Farm. By JACOB ABBOTT.

Shilling Toy Books. In demy 4to, stiff covers. With full-page Illustrations printed in Colours by KRONHEIM.

1. **Cock Sparrow.**
2. **Robinson Crusoe.**
3. **Queer Creatures.**
4. **Æsop's Fables.**

Shilling Reading Books.

Evenings at Home. In Words of One Syllable. Cloth limp, 1s.
Æsop's Fables. In Words of One Syllable. Cloth limp, 1s.

Eighteenpenny Series of New and Original Works. Bound in best cloth, gilt edges, with Four Coloured Plates by KRONHEIM in each Book. 128 pp., fcap. 8vo.

Little Blackcap. And other Stories.
Tommy and his Broom. And other Stories.
Little Red Shoes. And other Stories.
Charlie's Lessons about Animals.
The Broken Promise. By the Hon. Mrs. GREENE.
The Holidays at Llandudno.
The Hop Garden.
Algy's Lesson. By S. E. DE MORGAN.
Hid in a Cave.
Ashfield Farm.
Grandmamma's Spectacles. By the Author of "A Trap to Catch a Sunbeam."
Little Fables for Little Folks.
Flora Selwyn: A Story for Girls.

Two Shilling Series of New and Original Works. Bound in cloth gilt, gilt edges, with Illustrations printed in Colours. 160 pp., fcap. 8vo.

Dr. Savory's Tongs. By TWO SISTERS.
The Golden Gate. By H. G. B. HUNT.
Love and Duty. By ANNA J. BUCKLAND.
Brave Lisette. And other Stories. By Miss CARLESS.
Beatrice Langton; or, The Story of Ob[illegible]ience.
Owen Carstone: A Story of School Life.
New Stories and Old Legends. By Mrs. T. K. HERVEY.
The Little Orphan.
The Boy who Wondered.
The Story of Arthur Hunter and his First Shilling.
The Story of the Hamiltons.
The Hillside Farm.

Children's Books—(*continued*).

Half-Crown Library. 160 pp., extra fcap. 8vo, handsomely bound in cloth gilt, with Illustrations and Ornamental Chapter Heads.

Labour Stands on Golden Feet. A Holiday Story for the Industrial Classes. Translated from the German of HEINRICH ZSCHOKKE, by Dr. JOHN YEATS.

Stories of the Olden Time. Selected and Arranged by M. JONES.

One Trip More, and Other Stories. By the Author of "Mary Powell."

The Microscope, and some of the Wonders it Reveals. By the Rev. W. HOUGHTON. With numerous Illustrations.

Truly Noble: A Story. By Madame DE CHATELAIN.

Autobiographies of a Lump of Coal, a Grain of Salt, &c. By ANNIE CAREY.

The Fishing Girl. By BJOERNSTJERNE BJOERNSEN. Translated from the Norwegian by F. RICHARDSON and A. PLESNER.

Love and Life in Norway. By BJOERNSTJERNE BJOERNSEN. Translated from the Norwegian by the Hon. AUGUSTA BETHELL and A. PLESNER.

The Children's Library. A Series of Volumes specially prepared for Children. Beautifully Illustrated, and handsomely bound in cloth gilt, 224 pp., super-royal 16mo, uniform in size and price.

The Children's Sunday Album. By the Author of "A Trap to Catch a Sunbeam. With upwards of 100 Engravings. A Companion Volume to "The Children's Album," by UNCLE JOHN. Cloth gilt, 3s. 6d.

The Story of Robin Hood. Illustrated with Eight Plates printed in Colours. Cloth, 3s. 6d.

The True Robinson Crusoes. A Series of Stirring Adventures. Edited by CHARLES RUSSELL. Cloth gilt, 3s. 6d.

Off to Sea. By W. H. G. KINGSTON. With Eight Coloured Plates by KRONHEIM. Cloth gilt, 3s. 6d.

Old Burchell's Pocket. By ELIHU BURRITT. 3s. 6d.

The Children's Album. Containing Coloured Frontispiece and nearly 200 Engravings, with Stories by UNCLE JOHN, also several Pieces of Music. Twenty-sixth Thousand. 3s. 6d.

Peggy, and other Tales; including the HISTORY OF A THREEPENNY BIT, and THE STORY OF A SOVEREIGN, by the Author of "Misunderstood." 3s. 6d.

Mince-Pie Island: A Christmas Story for Young Readers. By R. ST. JOHN CORBET. 3s. 6d.

Cloudland and Shadowland; or, Rambles into Fairy Land with Uncle White Cloud. 3s. 6d.

The Queen of the Tournament. By the Author of "Mince-Pie Island." 3s. 6d.

Crocker the Clown: A Tale for Boys. By B. CLARKE. 3s. 6d.

Lily and Nannie at School: A Story for Girls. By the Author of "The Little Warringtons." 3s. 6d.

The Magic of Kindness. By the BROTHERS MAYHEW. With Eight Plates by WALTER CRANE. Cloth gilt, 3s. 6d.

On a Coral Reef: A Sea Story for Boys. By ARTHUR LOCKER. Cloth gilt, 3s. 6d.

King Gab's Story Bag; and the Wondrous Tales it contained. By HERACLITUS GREY. With Eight Plates by WALTER CRANE. 3s. 6d.

Hours of Sunshine: A Series of Poems for Children. By MATTHIAS BARR. With Sixteen Coloured Plates from Designs by OSCAR PLETSCH. 3s. 6d.

Playing Trades. By HERACLITUS GREY. With Sixteen Coloured Plates from Designs by J. BARFOOT. 3s. 6d.

The Angel of the Iceberg. And other Stories. By JOHN TODD, D.D. (New Edition.) 3s. 6d.

Drawing Room Plays. Gilt edges, 3s. 6d.

Famous Regiments of the British Army. By WILLIAM H. DAVENPORT ADAMS. Illustrated. 3s. 6d.

Will Adams: The Adventures of the First Englishman in Japan. By WILLIAM DALTON. 3s. 6d.

Working Women of this Century: The Lesson of their Lives. By CLARA LUCAS BALFOUR. 436 pp. 3s. 6d.

Dictionaries.

Bible Dictionary, Cassell's. With 600 Illustrations. One or Two Volumes, 21s.; bound in morocco, 40s.

Biographical Dictionary, Cassell's. 1,160 pp., imperial 8vo. Illustrated with Portraits. Cloth, 21s.; half-morocco or calf, 35s.

Brewer's Dictionary of Phrase and Fable; giving the Derivation, Source, or Origin of Common Phrases, Allusions, and Words that have a Tale to Tell. By the Rev. Dr. BREWER. Demy 8vo, 1,000 pp., cloth, 10s. 6d.

Cassell's Webster's Etymological Dictionary. 3s. 6d.

Dictionary of the English Language, A. 16th Edit., 3s. 6d.

French and English Dictionary. Crown 8vo, 956 pp., cloth, 3s. 6d.

German-English and English-German Pronouncing Dictionary. Crown 8vo, cloth, 864 pp., 3s. 6d.

Latin-English and English-Latin Dictionary. By J. R. BEARD, D.D., and C. BEARD, B.A. Cloth, 914 pp., 3s. 6d.

Illustrated National Dictionary, The, on the basis of Webster. With 250 Engravings. Crown 16mo, 1s.

Dictionary of Derivations, The. By Professor SULLIVAN, LL.D. 11th Edition. 2s.

Educational Works.

Algebra, Elements of. Paper covers, 1s.; cloth, 1s. 6d.

Anatomical Atlas, The. Designed and Printed in Colours from Nature. Royal folio, with Index, 10s. 6d.

Arithmetic.

Hudson's Arithmetic for School and College Use. With a copious Collection of Exercises and Key. 1s. 6d.

Wallace's Arithmetic. Cloth, 1s. 6d.

Arithmetic for the Young. Cloth, 1s.

Elementary Arithmetic. Part I., adapted to Standards I. and II. of the New Code. 64 pp., 4d. Key, 3d.

Elementary Arithmetic. Part II., adapted to Standards III. & IV. of the New Code. 80 pp., 6d. Key, 3d.

Elementary Arithmetic. Part III., adapted to Standards V. and VI. of the New Code. Cloth.

Book-keeping, by Single and Double Entry. 1s. Ruled Account Books to Ditto, extra, 1s. 6d. each Set.

Book-keeping for the Million. By THEODORE JONES. Cl., 3s.

Book-keeping for Schools. The English System. By THEODORE JONES. 2s.; cloth, 3s.

Book-keeping for Schools, The Key to. 2s.; cloth, 3s.

Books for Jones's System. Ruled Sets of. 2s.

Educational Works—(*continued*).

Brewer's Series of First Books. Price 6d. each.

Reading and Spelling.	Science.	Chemistry.
Bible History.	Common Things.	Facts and Discoveries.
History of England.	French History.	Grecian History.
Geography.	Astronomy.	The History of Rome.

Brewer's, The Young Tutor. First Series. Being the First Six Books in this Series, bound in One Volume. Cloth, 3s. 6d.

Brewer's, The Young Tutor. Second Series. Being the latter Six Books in this Series, bound in One Volume. Cloth, 3s. 6d.

Brewer's Guide to Every-day Knowledge. 284 pp., 2s. 6d.

Cassell's Technical Manuals, for Joiners, Carpenters, Machinists, Builders, Cabinet Makers, Stonemasons, Tin-plate Workers, Plumbers, and Artisans and Students generally.

Linear Drawing. By ELLIS A. DAVIDSON. With 150 Illustrations. 128 pp., extra fcap. 8vo, cloth limp, 2s.

Orthographic and Isometrical Projection. By the same Author. With 40 whole-page Diagrams. 128 pp., extra fcap. 8vo, cloth limp, 2s.

Linear Drawing and Projection. The Two Volumes in One. Cloth, lettered, 3s. 6d.

Building Construction, the Elements of, and Architectural Drawing. By ELLIS A. DAVIDSON. With 130 Illustrations. Extra fcap. 8vo, 128 pp., cloth limp, 2s.

Practical Perspective. By ELLIS A. DAVIDSON. With 36 double-page Illustrations. 90 pp., cloth, 3s.

Systematic Drawing and Shading. By CHARLES RYAN, Head Master, Leamington School of Art. With numerous Illustrations and Drawing Copies. 128 pp., extra fcap. 8vo, cloth limp, 2s.

Drawing for Carpenters and Joiners. By ELLIS A. DAVIDSON. With 253 Illustrations and Drawing Copies. 104 pp., extra fcap. 8vo, cloth, 3s. 6d.

Drawing for Machinists and Engineers. By ELLIS A. DAVIDSON. With over 200 Illustrations and Diagrams. Extra fcap. 8vo, cloth, 4s. 6d.

Model Drawing. With numerous Illustrations. 3s.

Cassell's Elementary Atlas. 16 Coloured Maps. Fcap. 4to, 6d.

Cassell's Preparatory Atlas. 16 Coloured Maps. Crown 4to, 6d.

Cassell's First School Atlas. Coloured Maps. Crown 4to, 1s.

Cassell's Handy Atlas. 24 Coloured Maps, and Index. Crown 8vo, cloth, 2s. 6d.

Cassell's Beginner's Atlas. 24 Coloured Maps, and Index. Crown 4to, cloth, 2s. 6d.

Cassell's Introductory Atlas. 18 Coloured Maps, and Index. Cloth, 3s. 6d.

Cassell's Select Atlas. 23 Coloured Maps, and Index. Imperial 8vo, cloth, 5s.

Cassell's Comprehensive Atlas. 42 Coloured Maps, and Index. 10s. 6d.

Chemistry. Specially adapted for the use of Self-Students. 1s.

Chemistry, Elementary. By the Rev. H. MARTYN HART. 290 pp., crown 8vo, cloth, 3s. 6d.

Educational Works—(*continued*).

Copy Books for Schools, Cassell's. With Set Copies on every page. Price 2d. each.

1. Initiatory Exercises.
2. Letters and Combinations.
3. Short Words.
4. Capitals.
5. Text Hand.
6. Text and Round.
7. Round Hand.
8. Round and Small Hands.
9. Small Hand.
10. Text, Round, & Small Hands.
11. Introduction to Ladies' Hand.
12. Ladies' Hand.
13. Commercial Sentences.
14. Figures.

*** These Books are now very largely used in Schools, and have been described as the *cheapest* and *best* ever published.

Drawing-Copy Books, Cassell's Penny. Each Penny Number consists of a Book of Sixteen Pages fcap. quarto, half of the page being occupied by the Drawing Copy, and the other half left blank, for the use of the Pupil to draw upon.

The First Grade Series comprises—

1. Right Line Forms.
2. Curved Line Forms.
3. Right Line and Curved Line Forms in Perspective.
4. Ornamental Forms.
5. Floral Forms, Fruit, &c.
6. Groups of Objects.

The Second Grade Series comprises—

7. Freehand—Elementary Forms of Plants.
8. Freehand—Advanced Studies of Plants.
9. Freehand—Conventional Ornament, principally based on Plant Forms.
10. Geometrical Problems.
11. Linear Perspective.
12. Model Drawing—Simple Objects of clearly-defined Forms.

These are followed by Books of Landscape and Marine Subjects, Birds, Dogs, and Miscellaneous Animals.

Drawing Copies, Cassell's Sixpenny.

Series A. **Floral and Vegetable Forms.** Twelve Parts, 6d. each; Twelve Packets on Cardboard, 1s. each.

Series B. **Model Drawing.** Twelve Parts, 6d. each; Twelve Packets on Cardboard, 1s. each.

Series C. **Landscape Drawing.** Twelve Parts, 6d. each; Twelve Packets on Cardboard, 1s. each.

Series D. **Figure Drawing.** Twelve Parts, 6d. each; Twelve Packets on Cardboard, 1s. each.

Series E. **Animal Drawing.** Twelve Parts, 6d. each; Twelve Packets on Cardboard, 1s. each.

The Drawing Copies, in 6d. Parts, and 1s. Packets on Cardboard, may be had in separate Parts or Packets.

Drawing Copies.

Gregory's First Grade Freehand Outline Drawing Examples. In Two Packets, each containing Twenty-four Examples, on Card, price 1s. per Packet. Enlarged for Black-board, 2s. 6d. per Packet.

Gregory's Outlines from Flowers. Twelve large Cards, in Packet, 1s. 6d.

Gregory's Outlines from Models used in Schools, on Twelve Cards, 1s.; and enlarged for Black-board, 2s.

Gregory's Easy Drawing Examples. Twenty-four Cards in Packet, 1s.; enlarged for Black-board, 2s. 6d.

English Grammar. By Professor SULLIVAN, LL.D. 1s.

Etymology, a Manual of. By Professor SULLIVAN. 10d.

English Spelling and Reading Book. With 160 Illustrations. 1s.

Euclid, Cassell's. Edited by Professor WALLACE, A.M., of the Glasgow University. 1s.; cloth, 1s. 6d. Key to Ditto, 4d.

Educational Works—(*continued*).

French, Cassell's Lessons in. Parts I. and II., in paper, each 2s.; cloth, each 2s. 6d. Complete in One Volume, 188 pp., 4s. 6d.

French, Key to the Exercises in Cassell's Lessons in. 1s.; Cloth, 1s. 6d.

French, Shilling Lessons in. By Professor DE LOLME. 1s.; cloth, 1s. 6d.

French, Sixpenny Lessons in. 6d.

French and English Correspondence for Boys. 2s. 6d.

French and English Correspondence for Young Ladies. 2s. 6d.

French and English Commercial Correspondence. 2s. 6d.

French Reader, The. New Edition. By Professor DE LOLME. Paper, 2s.; cloth, 2s. 6d.

Galbraith and Haughton's Scientific Manuals. Cloth, red edges.

Arithmetic. Cloth, 3s. 6d.
Plane Trigonometry. 2s. 6d.
Euclid. Elements I., II., III. 2s. 6d.
Euclid. Books IV., V., VI. 2s. 6d.
Mathematical Tables. 3s. 6d.
Mechanics. Cloth, lettered, 3s. 6d.
Optics. 2s. 6d.
Hydrostatics. 3s. 6d.
Astronomy. 5s.
Steam Engine. 3s. 6d.
Algebra. Third Edition. Part I., 2s. 6d.; complete, 7s. 6d.
Tides and Tidal Currents. New Edition, with Tidal Cards. 3s.
Natural Philosophy. With 160 Illustrations. 3s. 6d.
The Three Kingdoms of Nature. With 230 Illustrations. 5s.

Geography. By Professor SULLIVAN, LL.D. 1s.

Geography Generalised. By Professor SULLIVAN, LL.D. 2s.

German Reader, The International, for the Use of Colleges and Schools. By EDWARD A. OPPEN, of Haileybury College. 4s. 6d.

German, Lessons in. By W. H. WOODBURY. Parts I. and II., 2s.; cloth, each, 2s. 6d. Complete in One Volume, cloth, 4s. 6d.

German, Key to Lessons in. 1s.; cloth, 1s. 6d.

German, Sixpenny Lessons in. 6d.

The Literary Class Book. By Professor SULLIVAN. 2s. 6d.

Natural History of the Raw Materials of Commerce, The. *Second Edition.* By J. YEATS, LL.D., F.R.G.S. Intended for the Study of Young Merchants, Manufacturers, and Business Men. 452 pp., crown 8vo, cloth, 5s. (For "Technical History of Commerce," *see* p. 13.)

Natural Philosophy, in Easy Lessons. By Professor TYNDALL, F.R.S. Illustrated. New Edition, 2s. 6d.

Educational Works—*(continued)*.

Natural Philosophy, The Elements of, for the Use of Schools. By the Rev. SAMUEL HAUGHTON, M.D., F.R.S., Fellow of Trinity College, Dublin. With 160 Illustrations. Cloth, 3s. 6d.

Penny Table Book, Cassell's. For the Use of Schools. 1d.

Poetical Reader, Cassell's, for Pupil Teachers and School Use. 208 pp. Cloth, 1s.

Popular Education and School-keeping. By Professor SULLIVAN, LL.D. Second Edition. 2s.

Popular Educator, Cassell's New. Revised to the Present Date, with Numerous Additions. Vols. I., II., III., IV., V. and VI. now ready. Best cloth gilt, 6s. each; complete in Three Volumes, half-calf, £2 10s.

Primary Series. An entirely new and original Series of Volumes, specially prepared for the use of Elementary, National, and other Schools; written by men of practical experience.

Elementary Arithmetic. Simple Rules. 4d. Key to ditto, 3d.

Elementary Arithmetic. Compound Rules. 6d. Key to ditto, 3d.

Elementary British History. 6d.

Elementary Geography. 4d.

A Handy Book on Health, and How to Preserve it. Cloth, 9d.

England at Home. An Elementary Text-Book of Geography, Manufacture, Trade, and Commerce. 1s.

Our Bodies. An Elementary Text-Book of Human Physiology. 1s.

Our Houses, and what they are made of. 1s.

Our Food Supply. Elementary Lessons in Domestic Economy. 1s.

Our First Grammar. An Elementary Text-Book. 1s.

The Uses of Plants, in Food, Arts, and Commerce. With Illustrations. 1s.

Right Lines in their Right Places; or, Geometry without Instruments. With Drawings on Wood by the Author. 1s.

Vegetable Physiology, in Easy Lessons, with numerous Illustrations. 1s. 6d.

The Animal Kingdom. With abundant Illustrations. Double Vol. cloth, lettered, 200 pp., 2s.

CASSELL'S NEW CODE READERS,

Adapted to the Requirements of the New Code.

Elementary Boy's Reader, for Boys under Four. 4d.

Elementary Girl's Reader, for Girls under Four. 4d.

The Boy's First Reader. Standard I. 64 pp. Illustrated. Cloth, 4d.

The Girl's First Reader. Standard I. 64 pp. Illustrated. Cloth, 4d.

The Boy's and Girl's Second Reader. Standard II. 112 pp. Illustrated. Cloth, 6d.

The Boy's and Girl's Third Reader. Standard III. 128 pp. Cloth, 7d.

The Boy's and Girl's Fourth Reader. Standard IV. 160 pp. Cloth, 8d.

The Boy's and Girl's Fifth Reader. Standard V. 176 pp. Cloth, 10d.

The Boy's and Girl's Sixth Reader. Standard VI. 208 pp. Cloth, 1s.

Technical Educator, Cassell's. Vol. I., 412 pp., crown 4to, profusely Illustrated. Cloth, 6s.

The Spelling-Book Superseded. By Professor SULLIVAN. 1s. 4d.

Upwards of Four Hundred Words, Spelled in Two or more Ways; with an Attempt to settle their Orthography. By Professor SULLIVAN, LL.D. 10d.

Fine Art Volumes.

Æsop's Fables. With numerous Illustrations from Original Designs by ERNEST GRISET. Imperial 8vo, 236 pp., cloth, 7s. 6d.; full gilt, gilt edges, 10s. 6d.

After Ophir. By Captain A. F. LINDLEY. Illustrated with about Sixty Engravings. Cloth gilt, 7s. 6d.

Arms and Armour. By CHARLES BOUTELL, M.A., Author of "English Heraldry." With numerous Engravings. 7s. 6d.

Beauties of Poetry and Gems of Art. With Thirty-two Illustrations by J. C. HORSLEY, R.A., J. TENNIEL, C. W. COPE, R.A., PICKERSGILL, &c. With Ornamental Borders, &c. 7s. 6d.

Book of Historical Costumes, The. With Ninety-six full-page Coloured Engravings. 50s.

British Army, The History of the. By Sir SIBBALD DAVID SCOTT, Bart. Dedicated, by special permission, to the Queen. Two Volumes, 1,178 pp., demy 8vo, cloth, £2 2s.

Bunyan. The Pilgrim's Progress. Cloth, 7s. 6d.; gilt edges, 10s. 6d.; morocco antique, 21s.

Bunyan. The Holy War. Uniform with the above, and same price.

Chefs-d'œuvre of the Industrial Arts. With 200 Illustrations. By PHILIPPE BURTY. Edited by W. CHAFFERS, F.S.A. Cloth, 16s.; extra cloth gilt, £1 1s.

Crusoe, Life and Adventures of. With 100 Illustrations by MATT MORGAN, HARRISON WEIR, R. P. LEITCH, &c. Cloth, 7s. 6d.; full gilt, 10s. 6d.; morocco, 21s.

Doré Gallery, The. Containing 250 of the finest Drawings of GUSTAVE DORÉ. Letterpress and Memoir by EDMUND OLLIER. Cloth gilt, £5 5s.; full morocco elegant, £10; in Two Vols., cloth gilt, £5 10s.

Doré Bible. (See Bibles and Religious Literature.)

Doré's Milton's Paradise Lost. Illustrated with full-page Drawings by GUSTAVE DORÉ. Cloth, £5; best polished morocco, gilt extra, £10.

Doré's Dante's Inferno. Cloth, £2 10s.; morocco antique, with gilt edges, £4 4s.; full morocco, £6 6s.

Doré's Dante's Purgatory and Paradise. Uniform with the INFERNO, and same price.

Doré's Don Quixote. With 400 Illustrations. Cloth, £1 10s.; half morocco, £2 5s.; full morocco antique, £3 10s.

Doré's Atala. By CHATEAUBRIAND. Cloth, £2 2s.; morocco gilt, £4 4s.

Doré's La Fontaine's Fables. With Eighty-six full-page and many other Engravings. Cloth, £1 10s.; half-morocco, £2 5s.; full morocco antique, £3 10s.

Fine Art Volumes—(continued).

Doré's Fairy Realm. Illustrated with Twenty-five full-page Engravings by GUSTAVE DORÉ. Cloth gilt, gilt edges, £1 1s.

Doré's, The History of Croquemitaine, and the Times of Charlemagne. With nearly 200 Engravings. Cloth, 10s. 6d.

Doré's, The Adventures of Munchausen. With Thirty-one full-page Engravings. Cloth, 10s. 6d.

Doré's, The Legend of the Wandering Jew. Folio, 15s.; extra gilt, 21s.

English Heraldry. By CHARLES BOUTELL, M.A., Author of "Arms and Armour," &c. With 460 Engravings. New Edition. Cloth, gilt top, 5s.

Favourite Poems by Gifted Bards. Illustrated with Twenty-four Engravings by eminent Artists. Cloth, gilt edges, 7s. 6d.

Foxe's Book of Martyrs. Illustrated with 181 Engravings. Plain cloth, 12s.; full gilt cloth, gilt edges, 15s.

Goethe Gallery. A Series of beautiful Photographs from KAULBACH'S Drawings of the Heroines of Goethe. Morocco, 42s.

Goethe's Heroines. A Series of Twenty-one exquisite Engravings on Steel. Cloth, lettered, £7 7s.

Greece, The Scenery of. By W. LINTON. Fifty exquisitely beautiful full-page Steel Engravings. Cloth, lettered, gilt edges, 21s.

Gulliver's Travels. By DEAN SWIFT. With Eighty-eight Engravings by MORTEN. Imperial 8vo, 400 pages, plain cloth, 7s. 6d.; full gilt cloth, gilt edges, 10s. 6d.; full morocco antique, 21s.

Illustrated Readings. Vol. I., cloth gilt, 7s. 6d.; full gilt, gilt edges, 10s. 6d. Vol. II., cloth, 7s. 6d.; full gilt, gilt edges, 10s. 6d. Or, the Two Vols. in One, cloth, 12s. 6d.; half morocco, 15s.

Illustrated Travels. Edited by H. W. BATES, Assistant-Secretary of the Royal Geographical Society. Vols I. and II., each, cloth, 15s.; cloth, extra gilt, gilt edges, 18s. Or, the Two Vols. in One, cloth, 25s.; best cloth, gilt edges, 31s. 6d.

Insect World, The. From the French of LOUIS FIGUIER. With 570 Illustrations. Edited by E. W. JANSEN, Lib. E.S. Cloth, lettered, 16s.; extra cloth gilt, 21s.

Jewels Gathered from Painter and Poet. Cloth, gilt, 7s. 6d.

Log of the Fortuna, The. By Captain A. F. LINDLEY. Fcap. 4to, 256 pp., with 50 Engravings. Cloth gilt, 7s. 6d.

Millais' Illustrations. A Collection of Eighty Drawings on Wood, by JOHN EVERETT MILLAIS, R.A. Cloth, gilt edges, 16s.

Ocean World, The. From the French of LOUIS FIGUIER. Edited by C. O. G. NAPIER, F.G.S. Cloth, 16s.; extra cloth gilt, 21s.

Old Friends with New Faces. With Twenty-four full-page Illustrations, beautifully printed in Colours by KRONHEIM. 5s.

Fine Art Volumes—(*continued*).

Pictures from English Literature. With Twenty full-page Illustrations by J. C. HORSLEY, R.A., W. F. YEAMES, A.R.A., MARCUS STONE, J. GILBERT, H. K. BROWNE, W. CAVE THOMAS, LAWSON, HUGHES, BARNARD, FILDES, &c. &c. The Text by Dr. WALLER. Crown 4to, cloth gilt, 21s.

Reptiles and Birds. From the French of LOUIS FIGUIER. Edited by PARKER GILMORE, Esq. 18s. ; extra cloth gilt, 23s.

Sacred Poems, The Book of. Edited by the Rev. R. H. BAYNES, M.A. Cloth, 7s. 6d. ; extra gilt, gilt edges, 10s. 6d. ; morocco, 21s.

Schiller Gallery, The. A Series of choice Photographs from KAULBACH's Paintings of Scenery from SCHILLER. £5 5s.

Selection of One Hundred of the Finest Engravings by the late G. H. Thomas. Cloth gilt, 10s. 6d.

Thorwaldsen's Triumphal Entry of Alexander the Great into Babylon. Twenty-two Plates, folio. 42s.

Vegetable World, The. With 471 Illustrations. From the French of LOUIS FIGUIER. Edited by C. O. G. NAPIER, F.G.S. Cloth, lettered, 16s. ; extra cloth gilt, 21s.

Vicar of Wakefield, The, and POEMS. Beautifully printed on Toned Paper, and Illustrated with 108 Engravings. In one handsome Imperial 8vo Volume, 378 pp., bound in cloth, 7s. 6d. ; full gilt cloth, with gilt edges, 10s. 6d. ; full morocco antique, 21s.

World before the Deluge, The. With 233 Illustrations. From the French of LOUIS FIGUIER. Edited by H. W. BRISTOW, F.R.S. Cloth, lettered, 16s. ; extra cloth gilt, 21s.

World of the Sea. With 18 Coloured Plates and numerous Wood Engravings. Cloth, lettered, 21s.

Hand-Books and Guides.

Bacon's Guide to America and the Colonies. With Maps, &c. Cloth, 2s. 6d.

Civil Service, Guide to Employment in the. Revised to the Present Time. Cloth, 2s. 6d.

Civil Service, Guide to the Indian. By A. C. EWALD, F.S.A. Cloth, 4s. 6d.

Guide to America for the Capitalist, Tourist, or Emigrant. By G. W. BACON, F.R.S. With Coloured Map, 1s.

Emigrant's Guide to the Colonies of Great Britain. 6d.

Household Guide, The.

Vols. I., II., and III., each containing Two Coloured Plates, and Illustrations on nearly every page. Cloth gilt, 6s. per volume.

Hand-Books and Guides—(*continued*).

Hand-Books, Cassell's Popular. Cloth, 1s. each; free by post for 13 stamps.

- **Art of Conversation.**
- **Book-Keeping,** by Single and Double Entry. Ruled Account Books to ditto, extra, 1s. 6d. each Set.
- **Business.**
- **Chemistry.**
- **Chess and Draughts.**
- **Domestic Pets.**
- **Domestic Recipes.**
- **Drawing-Room Magic.**
- **Elocution and Oratory.**
- **Emergencies.**
- **Etiquette** for Ladies and Gentlemen.
- **Gardening.**
- **How to Colour a Photograph** in Oil or Water.
- **Investments.**
- **Letter-Writing.**
- **Natural Philosophy.**
- **Photography.**
- **Railway Situations.**

Thames and Tweed.—New Work on Fishing. By GEORGE ROOPER, Author of "Flood, Field, and Forest." 2s. 6d.

History.

England, Illustrated History of. Complete in Eight Volumes, bound in cloth, 6s. and 7s. 6d. each. Ditto, Four Volumes, strongly bound in half-calf, with full gilt backs and cloth sides, £4.

———— The Toned Paper Edition, Vols. I., II., III., IV., and V. Cloth, each, 9s.

England, History of, from the First Invasion by the Romans to the Accession of William and Mary in 1688. By JOHN LINGARD, D.D. In Ten Volumes. Cloth, lettered, 35s.

Technical History of Commerce. By J. YATES, LL.D., F.R.G.S. Cloth, 5s. (For "Natural History of Commerce," *see* p. 8.)

Miscellaneous.

A Poet-Hero. By COUNTESS VON BOTHMER. Being a Biography of the German War-Poet, Theodore Körner. Cloth, 6s.

Appropriation of the Railways by the State. By ARTHUR JOHN WILLIAMS, Barrister-at-Law. The People's Edition. Crown 8vo, 1s.

Belle Sauvage Library, The. Price 3s. 6d. per Vol., cloth, fcap. 8vo.

1. **Pulpit Table Talk.**
2. **The Search for the Gral.** By JULIA GODDARD.
3. **Sermons for Boys.** By the Rev. ALFRED BARRY, D.D., Principal of King's College.
4. **The Life of Bernard Palissy, of Saintes.** By HENRY MORLEY.
5. **The Young Man in the Battle of Life.**

Burritt, Elihu. Thoughts and Notes at Home and Abroad. 6s.

Cassell's Magazine.

Vol. I., New Series, containing "Man and Wife," by WILKIE COLLINS. 600 pp. of Letterpress and Engravings. Cloth, gilt lettered, 6s. 6d.

Vol. II., containing Mr. J. S. LE FANU's New Story, "Checkmate," complete. Profusely Illustrated. Cloth, 5s. 6d.

Cameron's Handy Book of Food and Diet. Cloth, 1s.

Miscellaneous Works—(*continued*).

Daybreak in Spain. By the Rev. Dr. WYLIE. With Twelve Illustrations. 6s.

Flood, Field, and Forest. By G. ROOPER. With Fifteen Illustrations. Cloth, 3s. 6d.

History of the Pianoforte. By EDGAR BRINSMEAD. Illustrated. Cloth, 3s.

Little Gem Series, The. Cloth, 6d. each; red edges, 9d. each.

- **Shall we Know One Another?** By the Rev. J. C. RYLE
- **The Grounded Staff.** By the Rev. R. MAGUIRE.
- **Words of Help for Everyday Life.** By Rev. W. M. STATHAM.
- **The Voice of Time.** By J. STROUD.
- **Pre-Calvary Martyrs.** By the Rev. J. B. OWEN, M.A.
- **Home Religion.** By the late Rev. W. B. MACKENZIE, M.A.

North-West Passage by Land, The. By Viscount MILTON, M.P., and W. B. CHEADLE, B.A. 21s. Ditto, ditto, Smaller Edition, complete, 6s.

Penny Library of Popular Authors.

1. **Foxe's Book of Martyrs.** Price 1d.
2. **Bunyan's Pilgrim's Progress.** Price 1d.

Practical Poultry Keeper, The. By L. WRIGHT. *Fourth Edition.* With Thirty-six Plain Illustrations. Cloth, 3s. 6d.; with Twelve Coloured ditto, ditto, crown 8vo, 5s.

Quiz, Sketches by. Illustrated by "PHIZ." Cloth, 3s. 6d.

Ready Reckoner, Cassell's Sixpenny. Cloth, 6d.

Ready Reckoner, Cassell's Shilling. Containing Calculations from $\frac{1}{16}$ of 1d. to £1, Interest, Profit, and Commission Tables, &c. &c.

Romance of Trade. By H. R. FOX BOURNE. Cloth, 5s.

San Juan Water Boundary Question. By Viscount MILTON, M.P. Cloth, lettered, 10s. 6d.

Woman; Her Position and Power. By W. LANDELS, D.D. Cloth, lettered, 3s. 6d.

Wonders, Library of. Fully Illustrated. A Series of Gift Books and School Rewards. Cloth gilt, gilt edges, each 5s.

- **Wonders of Animal Instinct.**
- **Wonders of Bodily Strength and Skill.**
- **Wonders in Acoustics.**
- **Wonderful Balloon Ascents.**
- **Wonderful Escapes.**
- **Wonders of Architecture.** With Fifty-four Illustrations.

World of Wonders, The. With 130 Illustrations. Cloth, 7s. 6d.; full gilt, 10s. 6d.

Natural History.

Book of Birds, Cassell's Brehm's. Translated from the Text of Dr. BREHM by T. RYMER JONES, F.R.S. Vol. I. contains Ten Coloured Plates and 384 pp. letterpress, with numerous Illustrations. Cloth, 7s. 6d.; extra gilt, 10s. 6d.

Natural History, Cassell's Popular. Two vols., cloth, 30s.; half-calf, full gilt back, 45s.; half-morocco, full gilt, 50s. Ditto, with Coloured Illustrations, Four Volumes, cloth, 42s.

Natural History, Picture. Edited by the Rev. C. BOUTELL, M.A. With 600 Illustrations. Cloth, lettered, 3s. 6d.

Popular Scientific Library, The.

The World before the Deluge. With 233 Illustrations. Cloth, lettered, 16s.; extra cloth gilt, £1 1s.

The Vegetable World. With 471 Illustrations. Cloth, lettered, 16s.; extra cloth gilt, £1 1s.

The Ocean World. With 427 Illustrations. Cloth, lettered, 16s.; extra cloth gilt, £1 1s.

World of the Sea. Translated from the French of MOQUIN TANDON, by the Rev. H. M. HART. With Eighteen Coloured and Tinted Plates, and numerous Wood Engravings. Best cl., lettered, 21s.

The Insect World. With 576 Illustrations. Cloth, lettered, 16s.; extra cloth gilt, £1 1s.

Birds and Reptiles. With 307 Illustrations. Cloth, lettered, 18s.; extra cloth gilt, £1 3s.

Transformations of Insects. Translated and Adapted by Dr. DUNCAN, Secretary of the Geological Society, and Professor of Geology, King's College, London, from the French of EMILE BLANCHARD. With 240 highly-finished Engravings. Cloth gilt, gilt edges, 16s. *Second Edition.*

Poetry and the Drama.

Barr's Poems. New and Revised Edition. Cloth, 5s.

Book of Sacred Poems. Illustrated. Edited by the Rev. R. H. BAYNES, M.A. Cloth, 7s. 6d.; full gilt, 10s. 6d.; morocco, 21s.

Bright Thoughts for the Little Ones. With Twenty-seven Original Drawings by PROCTER. Cloth, lettered, gilt edges, 7s. 6d.

Cassell's Library Edition of the British Poets. Vols. I. and II. now ready, price 2s. each.

Child's Garland of Little Poems, The. With Exquisite Illustrative Borders by GIACOMELLI. Cloth gilt, 7s. 6d.

Favourite Poems by Gifted Bards. With Twenty-four Illustrations. Cloth, gilt edges, 7s. 6d.

Golden Leisures. By W. GORDON SMYTHIES. 1s.

Greece, The Poets of. By EDWIN ARNOLD, M.A., Oxon; Author of "Griselda and other Poems," &c. Demy 8vo, 256 pages, 5s.

Hours of Sunshine. By MATTHIAS BARR, Author of "Little Willie," &c. With Sixteen Coloured Plates. Cloth gilt, 3s. 6d.

Jewels Gathered from Painter and Poet. Cl., gilt edges, 7s. 6d.

Poets, Cassell's Three-and-Sixpenny Editions of the. In fcap. 8vo, printed on Toned Paper, elegantly bound in cloth, extra gold and colours, 3s. 6d. each: best morocco, inlaid with enamel letter-piece, 6s. 6d. each. LONGFELLOW; SCOTT; BYRON; MOORE; WORDSWORTH; COWPER; MILTON; POPE; BURNS; THE CASKET OF GEMS; THE BOOK OF HUMOROUS POETRY; BALLADS, SCOTTISH AND ENGLISH; LIVES OF THE BRITISH POETS.

Poetry and the Drama—(*continued*).

Poems and Pictures. With 100 Illustrations. Cloth gilt, gilt edges, 21s.

Shakespeare, Cassell's Illustrated. With 500 Illustrations. Imp. 8vo. Edited by CHARLES and MARY COWDEN CLARKE. Vol. I. (COMEDIES), 12s.; Vol. II. (HISTORICAL PLAYS), 10s. 6d.; Vol. III. (TRAGEDIES), 12s. 6d. The Complete Work, in Three Volumes, cloth, lettered, uniform, £1 15s. Half morocco, £2 10s. The Separate Plays may be had, price 1s. each.

Serial Publications.

Cassell's Popular Natural History. Monthly, 6d.; Weekly, 1d.

Little Folks. Monthly, 6d.; Weekly, 1d.

Cassell's History of the War between France and Germany. Monthly, 7d.; Weekly, 1½d.

Cassell's Penny Drawing-Copy Books. Weekly, 1d.

Book of Birds. Monthly, 7d.

Cassell's Magazine. Monthly, 6d.; Weekly, 1d.

British Poets. Monthly, 6d.

Family Bible. Monthly, 7d.; Weekly, 1½d.

Goldsmith's Works. Monthly, 5d. and 6d.; Weekly, 1½d.

Doré Milton. Monthly, 2s.

Doré Don Quixote. Monthly, 7d.; Weekly, 1½d.

Household Guide. Monthly, 7d.; Weekly, 1½d.

The Quiver. Monthly, 6d.; Weekly, 1d.

Child's Book of Song and Praise. Monthly, 6d.; Weekly, 1d.

Matthew Henry's Commentary. Monthly, 7d.; Weekly, 1½d.

Technical Educator. Monthly, 7d.; Weekly, 1½d.

Illustrated Travels. Monthly, 1s.

History of England. Monthly, 7d.; Weekly, 1½d.

Railway Time Tables and Trades' Directory. Monthly, 2d.

☞ For detailed particulars of all the Works inserted in the foregoing List consult CASSELL, PETTER, and GALPIN'S DESCRIPTIVE CATALOGUE, supplied by all Booksellers, and forwarded Post Free by the Publishers.

www.ingramcontent.com/pod-product-compliance
Lightning Source LLC
Chambersburg PA
CBHW021512210326
41599CB00012B/1226

9783741127069